# A Primer on Earth Pollution: Pollution Types and Disposal

Edited by

## J. Senthil Kumar

*Department of Biotechnology,
Sri Krishna Arts and Science College,
Coimbatore, Tamil Nadu,
India*

## P. Ponmurugan

*Department of Botany,
Bharathiar University,Coimbatore,
Tamil Nadu,
India*

&

## A. Vinoth Kanna

*Department of Biotechnology,
Sri Krishna Arts and Science College,
Coimbatore, Tamil Nadu,
India*

# A Primer on Earth Pollution: Pollution Types and Disposal

Editors: J. Senthil Kumar, P. Ponmurugan, A. Vinoth Kanna

ISBN (Online): 978-981-14-7655-6

ISBN (Print): 978-981-14-7653-2

ISBN (Paperback): 978-981-14-7654-9

Published by Bentham Science Publishers Pte. Ltd. Singapore. All Rights Reserved.

need for a court order if at any point you breach any terms of this License Agreement. In no event will any delay or failure by Bentham Science Publishers in enforcing your compliance with this License Agreement constitute a waiver of any of its rights.

3. You acknowledge that you have read this License Agreement, and agree to be bound by its terms and conditions. To the extent that any other terms and conditions presented on any website of Bentham Science Publishers conflict with, or are inconsistent with, the terms and conditions set out in this License Agreement, you acknowledge that the terms and conditions set out in this License Agreement shall prevail.

**Bentham Science Publishers Pte. Ltd.**
80 Robinson Road #02-00
Singapore 068898
Singapore
Email: subscriptions@benthamscience.net

BENTHAM
SCIENCE

# CONTENTS

# PREFACE

Pollution is one of the serious cases that need to be taken care of. Pollution can be broadly classified as air, land, and water pollution. The term defines the undesirable changes in the specific environmental condition. The study on pollution is one of the extensive areas of study in almost all countries around the Globe. The byproducts and used over materials of like plastics, one use plastic bags, medical wastes and other pollutants were reported since two decades with their adverse effects.

Pollution as a whole in any form should either limited or reduced in order to gift the wealth we enjoyed from our mother nature. This book is in lieu with various pollution and the problems associated with us in our day to day life. It comprises of 13 chapters including Agricultural pollution, Land Pollution, Water Pollution, Biomedical waste pollution, Micro plastics pollution and Synthetic dyes pollution, *etc*. The book also highlights the readers with the damage incurred to human health by the special chemical substances present in air, water , food and radioactive compounds. This will create adverse effects such as plants, animals, vegetation and vegetation. The damage caused for long term will leads to an apparent and embracing situation may not fit for humans to live.

It educates readers with about the probable arenas of pollution and ways to control it. Paws a path before we leap and we should return all the wealth for the next generation to grow and survive better with healthy atmosphere. The reductions of pollution in all means are in its way, as one among the Hundred billion question of the hour.

## ACKNOWLEDGEMENTS

The Editors of this book are thankful to the Management of Sri Krishna Arts and Science College, Coimbatore, and Bharathiar University, Coimbatore India, for providing an opportunity to publish this book under the umbrella of Bentham Science Publishers, Singapore.

The Editors are also thankful to the Science Academies, India, for their appreciation and motivation of the participants in connection with the collaborative teamwork required to make this book a success.

The Editors are also thankful to the contributors for their extended support in providing their chapters within the stipulated time and in an effective way to allow for faster processing.

The Editors are thankful for to, for accepting our proposal and assisting us Bentham Science Publishers, Singapore through their expertise and their financial support.

The Editor also appreciates Nature Science Foundation, Coimbatore for their motivation and their assistance during the processing of this book.

## CONSENT FOR PUBLICATION

Not applicable.

## CONFLICT OF INTEREST

The authors confirm that this chapter contents have no conflict of interest.

**J. Senthil Kumar**
Department of Biotechnology
Sri Krishna Arts and Science College
Coimbatore, Tamil Nadu
India

# LIST OF CONTRIBUTORS

| | |
|---|---|
| **A. Anitha** | Department of Biotechnology, Nehru Arts and Science College, Tamil Nadu 641105, India |
| **B. Preetham Kumar** | Department of Biotechnology, Sri Krishna Arts and Science College, Coimbatore, Tamil Nadu, India |
| **B. Sathya Priya** | Department of Environmental Sciences, Bharathiar University, Coimbatore-641046, Tamil Nadu 641046, India |
| **Gnanendra Shanmugam** | Department of Biotechnology, Yeungnam University, Gyeongsangbuk-do, South Korea |
| **J. Beslin Joshi** | Department of Plant Biotechnology, Tamil Nadu Agricultural University, Coimbatore, Tamil Nadu, India |
| **J. Senthil Kumar** | Department of Biotechnology, Sri Krishna Arts and Science College, Coimbatore, Tamil Nadu, India |
| **K. Chitra** | Department of Botany, Bharathiar University, Coimbatore, Tamil Nadu, India |
| **M. Sudha Devi** | Department of Biochemistry, Avinashilingam Institute for Home Science and Higher Education for Women, Coimbatore, Tamil Nadu, India |
| **M. Suguna Devakumari** | Department of Agriculture, Karunya Institute of Technology and Sciences, Coimbatore, Tamil Nadu, India |
| **Michelle Fong Ting Lim** | Department of Biomedical Sciences, MAHSA University, Selangor, Malaysia |
| **Nagaraja Suryadevara** | Department of Biomedical Sciences, MAHSA University, Selangor, Malaysia |
| **P. Ponmurugan** | Department of Botany, Bharathiar University, Coimbatore, Tamil Nadu, India |
| **R. Akshaya** | Department of Biochemistry, Biotechnology and Bioinformatics, Avinashilingam Institute for Home Science and Higher Education for Women, Tamil Nadu 641043, India |
| **R. Padma** | Department of Biochemistry, Avinashilingam institute for Home Science and Higher Education for Women, Coimbatore, Tamil Nadu, India |
| **R. Gomathi** | Department of Biotechnology, Sri Krishna Arts and Science College, Coimbatore, Tamil Nadu, India |
| **R. Sathya** | Department of Information Technology, Sri Krishna Adithya College of Arts and Science, Coimbatore, Tamil Nadu, India |
| **R. Vishnu** | Department of Biotechnology, Sri Krishna Arts and Science College, Coimbatore, Tamil Nadu, India |
| **S. Geethalakshmi** | Department of Biotechnology, Sree Narayana Guru College, Coimbatore, Tamil Nadu, India |
| **S. Sumathi** | Department of Biotechnology, Avinashilingam Institute for Home Science and Higher Education for Women, Coimbatore, Tamil Nadu, India |
| **T. Stalin** | Forestry Research and Development Unit, Molecular Biology Division, Karur-639 136, Tamil Nadu, India |
| **T. Dhanalakshimi** | Department of Biology, Ministry of Education, Coimbatore, Tamil Nadu, Republic of Maldives |

**V. Arun**                  Department of Biotechnology, Sri Ramachandra Institute of Higher Education and Research (SRIHER) (DU), Chennai - 600 116, Tamil Nadu, India

# CHAPTER 1

# Synthetic Dyes Pollution

**B. Sathya Priya**[1,*], **V. Arun**[2] and **T. Stalin**[3]

[1] *Department of Environmental Sciences, Bharathiar University, Coimbatore - 641 046, Tamil Nadu, India*

[2] *Department of Biotechnology, Sri Ramachandra Institute of Higher Education and Research (SRIHER) (DU), Chennai - 600 116, Tamil Nadu, India*

[3] *Forestry Research and Development Unit, Molecular Biology Division, Karur-639 136, Tamil Nadu, India*

**Abstract:** Colored products are more attractive and increase the marketing value. Natural and synthetic dyes are used for many centuries for coloring. Synthetic dyes are preferred mostly due to their high stability and cost effectiveness. The food products and textile fabrics which are coloured by synthetic dyes cause severe health issues to human beings. The synthetic dyes are cleaved into aromatic or aryl amines during reductive reactions that cause carcinogenic and mutagenic effects. Children are severely affected due to the consumption of artificial colours in food and may get ADHD (Attention Deficit Hyperactivity Disorder). The European Commission (EU) prohibits the marketing of products that contain the restricted azo dyes which have longer contact with the skin. The synthetic dyes with improper fixing are discharged into the environment along with effluent, causing biomagnification problems. They also affect the terrestrial and aquatic systems and cause severe pollution to the environment. This study deals with the various types of textile and food dyes, their impacts on health and environment and the effective treatment methods for the removal of dyes from the industrial effluent.

**Keywords:** Azo dyes, Biosorption, Biodegradation, Carcinogenic, Food Colourants, Synthetic Dyes.

## INTRODUCTION

The coloured fabrics or food products are liked by everyone, from the kids to adults. Natural dyes are more ecofriendly but not used more because of their less availability, low colour fastness, restricted colours, more cost needed for production and extraction. Hence, synthetic dyes are increasingly used by industries for colouring their products.

* **Corresponding author B. Sathya Priya:** Department of Environmental Sciences, Bharathiar University, Coimbatore - 641 046, Tamil Nadu, India; Tel: 8825688512; E-mail: sbspriya11@gmail.com

**J. Senthil Kumar, P. Ponmurugan & A. Vinothkanna (Eds.)**

Synthetic dyes are derived from carcinogenic petrochemical compounds and have more advantages such as less cost, varieties of colors, simple production with high colour-fastness. Azo dyes are produced in a larger quantity than other dyes and are mostly utilized by textile, leather and food industries. They cause many health disorders, including irritation of the skin, eye, lungs, mutagenic and cancer effects on human beings. Moreover, the mordant like chromium which is used to fix colour on the fabric, is also highly toxic and results in detrimental effects. The workers in the textile and dyeing industries are easily prone to cancer risks. The dyes are persistent in the environment as xenobiotics and are very difficult to treat by the conventional methods due to their complex structure. The textile and dyeing industries discharge a large quantity of wastewater, which has a significant level of dyes and other toxicants that have severe detrimental effects on the environment.

## SOURCES OF SYNTHETIC DYES POLLUTION

Synthetic dyes are used by many industries, such as textile, dyeing, paint, cosmetics, food, pharmaceutical, leather and plastics,for colouring the products. The discharge of the untreated effluent into the environment causes severe pollution problems. Textile industries use more azo dyes for processing in which some portion of the dyes is not fixed to the fabrics and washed out along with the effluent.

## Dyeing Process

Dyeing is the process of colouring the textile material with other treatments. Dyes contain both chromophoric and auxochromic groups. The chromophores and auxochromes are responsible for colouring and intensifying the colours. The batch, continuous and semi-continuous mode is used for the dyeing process. The batch mode is usually used for dyeing the fabrics. The aqueous solution is used to fix the dye on the fabric. During dye fixation, 4 distinctive forms of interaction may be observed, such as Van der Waals, ionic, hydrogen interactions and covalent bonds [1]. The process of dyeing has three important steps such as preparation, dyeing and finishing, which are as follows:

### *Preparation of Fabrics*

Before the dyeing process, the fabric is treated with alkaline substances, detergents and enzymes for removing the impurities. The fabrics are bleached with chlorine substances or hydrogen peroxide in order to remove the natural colour of the fabrics. Brightening agents are added to improve the white colour of the fabric.

## Dyeing

This step is responsible for the application of colour to the fabrics by the synthetic dyes due to diffusion and adsorption at high temperatures and pressures with the help of chemical aids such as acids, bases, carriers, surfactants, promoting, chelating agents, softening agents, *etc.* in order to get the uniform color and color fastness. Dyeing can also be done by using pigments with binders such as polymers, which fix the pigments to the respective fibers [2 - 5].

## Finishing

This is the final stage in which pressing, water proofing, softening and applying antimicrobial agents are used to enhance the quality of the fabric [5].

## TYPES OF DYES

Table **1** describes the different types of dyes with specific features [1, 4] and applications.

**Table 1. Different types of dyes.**

| Name of the Dyes | Characteristics and Application of Dyes | Examples |
|---|---|---|
| **Direct dye** Salt dyes or cotton colours | a. It is directly applied to fabrics with salts for fixation. b. Mostly applicable to cotton and also used for dyeing of linen, wool, silk, nylon and rayon. c. It has less fastness to light and washing. | Direct orange 26 |
| **Acid dye** anionic dyes | a. It is mostly used in acidic conditions. b. Wool and silk are coloured by this dye. c. It has high fastness to light and less to washing. | Acid Yellow 36. |
| **Basic dye** | a. It is used for dyeing wool, silk, linen and acrylic fibres without using mordant b. Cotton and rayon are dyed with a mordant tannic acid. c. Leather and paper industries also used this for dyeing. d. It gives bright shades to fabrics. e. It has less fastness but is enhanced by steaming. | Basic Brown 1 |
| **Azoic dye** Ice colours | a. Cotton, nylon and paper are coloured by this dye. b. It gives bright colours. c. Mostly ice or cold water is used in the dyeing process. d. It has high fastness to wash and light. | Bluish red azoic dye |
| **Vat dye** | a. Cotton, silk, wool and nylon are dyed. b. It has high fastness to light and wash. | Synthetic indigo, Vat Blue 4 |
| **Reactive dye** | a. Both hot and cold water reactive dyes are available. b. It is used for dyeing cotton, wool and silk fabrics. c. It has high fastness to light and wash. | Reactive Blue 5 |

(Table 1) cont.....

| Name of the Dyes | Characteristics and Application of Dyes | Examples |
|---|---|---|
| **Disperse dye** | a. Cellulose diacetate, cellulose triacetate and polyester fibers are dyed.<br>b. It gives bright shades to the fabric.<br>c. Mostly needs a carrier and dyed in hot condition.<br>d. It is also used for printing and mostly applicable to synthetic fabrics.<br>e. It has high fastness to light and wash. | Disperse Red 4 |
| **Sulfur dye** | a. It gives strong shades to cotton, rayon and linen.<br>b. It has high fastness to wash.<br>c. It is applied to cotton by using sodium sulfide as the reducing agent. | Sulfur Red 7 |
| **Nitro dye** | a. Wool is dyed by using this dye.<br>b. It has two or more aromatic rings. | Maritus yellow |
| **Mordant dye** | a. It needs a binding agent mordant.<br>b. Cotton and wool are dyed with this dye.<br>c. It has high fastness to wash and light. | Mordant Red 11 |
| **Solvent dye** | a. It is used for coloring the synthetics, plastics and waxes. | Solvent Yellow32 |

## VARIOUS TYPES OF FOOD COLORANTS

The food colorants are either natural or synthetic dyes, which are used to give attractive colour to the food products. Natural food dyes are safe for health but have limitations such as less colour, more cost, less availability and tedious extraction. So synthetic food dyes are mostly preferred due to colour stability and less cost. Children are more attracted to coloured foods than adults.

Table **2** shows the different types of food colours and their applications [6 - 12].

**Table 2. Different types of dyes used as food colourants.**

| Name of the Dye | Colour | Applications |
|---|---|---|
| **Rhodamine**<br>Basic dye | Pink | Colouring the food products. |
| **Alluran red (Red 40)**<br>Azo dye | Red | Used in candies, syrups, cosmetics, beverages and confectionaries. |
| **Erythrosine (Red3)**<br>Organo iodide compound | Cherry pink | Used in oral medication, candies, sausages and baked items. |
| **Lead chromate (Yellow)** or chrome yellow | Bright yellow | Mixed with turmeric powder to enhance colour. |
| **Green (fast green 3)** | Greenish | Used in cosmetics, drugs, candies and ice cream. |
| **Name of the Dye** | Colour | Applications |

*(Table 2) cont.....*

| Name of the Dye | Colour | Applications |
|---|---|---|
| **Food Brown** **Azo dye** | Brown | Used in fish products. |
| **Sunset Yellow (Yellow 6)** **Azo dye** | Yellow | Used in drugs, sausages, gelatine desserts and confectionery. |
| **Indigo carmine** **azo dye** | Deep blue and violet | Used in candies, beverages and pet food. |
| **Tartrazine (Yellow5)** | Yellow | Candies, bakery goods, pet foods, beverages, dessert powders, cosmetics, pharmaceuticals, gelatin desserts and cereals. |

EU (European Union) permits the use of 16 synthetic dyes. The dyes with brown and black colour are banned in most of the developed countries. The permissible limit of dye to any food item is 0.1 g/Kg [13].

## THE IMPACTS OF DYES

China and India produce maximum azo dyes. Azo dyes are mostly used because they need 60°C for dyeing, but the other azo-free dyes need 100°C. Moreover, they give an extensive range of colours, good colour fastness during synthesis and are cost-effective. But under reductive conditions, they cleave and produce dangerous aromatic amines, which are carcinogens in humans. They are traced in the dyed product and in the environment due to improper degradation.

### Environmental Impact of Dyes

The textile industries are one of the main polluters due to the discharge of effluent with toxic dyes and chemicals. The traces of azo dye in water affect the transparency, light penetration, photosynthetic activity, solubility of gas, depletion of oxygen. The azo dyes cause carcinogenic and mutagenic effects on organisms. They affect the yield of plants and cause metabolic disorders, damage to neurosensors, stress and death to aquatic organisms. They also affect the water used for drinking, recreation, aquaculture and irrigation. The azo dyes may alter the physical and chemical properties of soil and cause death to the soil microbes that affect the agricultural yield [14]. The improper dyeing process results in the release of dyes (10 -50%) along with the effluents and reaches the environment. The synthetic dyes are highly stable and persist as xenobiotics in the environment for long days. Dyes such as reactive, disperse and vat dyes are also responsible for the formation of AOX (Adsorbable Organic Halogen). Metals which are present along with dye stuffs cause many health disorders.

## Impacts of Food Dyes on Health

Amaranth may affect the immune system. Sunset yellow (Yellow 6) causes breast cancer. Auramine (yellow dye) and rhodamine cause damages to the kidney, liver and retard the growth condition. Rhodamine may breakdown red blood cells and Lead chromate causes anaemia, neurological disorders and hypertension. Metanil yellow severely affects the reproductive organs (ovaries and testis). It causes discoloration of the skin and may cause methaemoglobinaemia in adults after consuming rice coloured by it. Brown dye and tartrazine elevate asthma. Allura red and brown dyes may cause allergic reactions [15]. ADHD is an attention deficit hyperactivity disorder which may cause hyperactivity and lack of attention. The persons who regularly consume artificial food colours may get ADHD. Experimental studies showed that the children who consumed Yellow 5, Tartrazine and allura red regularly had more hypersensitivity reactions [16 - 18]. Red 40, Yellow 5 and 6 have benzidene, which is a carcinogen. Azo dyes follow three mechanisms to produce such a carcinogenic product. The reactive intermediate molecules are formed during metabolism that linked closely with the DNA [19]. The azo dyes, which are metabolized by the intestinal microflora, produce aromatic amines that cause cancer in the intestine [20]. The food and textile industries mostly use azo dyes and increase the cancer risk in people in the industrial area who are exposed to those dyes. Auramine and malachite green may cause mutagenic effects in the exposed organisms [21].

## Health Effects of Textile Dyes

The azo dyes have one or more nitrogen-nitrogen double bonds (azo groups). They are cleaved into aromatic amines, which are carcinogenic and mutagenic under reductive conditions that cause harmful health disorders to human beings. Azo dyes are mainly used for yellow, orange, and red coloring [22, 23]. The improper fixing of dyes is responsible for many health disorders [24]. The non-fixed dyes, which are nondegradable, are released along with effluent during discharge into the nearby environment [25]. Besides dyes, the wastewater also contains dyeing additives like ethoxylates, alkylphenol ethoxylates, retarders for cationic dyes, dispersing agents, ethylenediamine tetraacetate and many others [26]. Most of the azo dyes and other types of textile dyes such as anilines and anthraquinones are considered carcinogenic or mutagenic. Industrial effluents with azo dyes can cause mutagenic and genotoxic effects in cultured cells [27 - 29]. The direct or acid application type and fragrances of azo dyes can cause potential risks such as cancer and allergy to human health if they are exposed. It may be caused by inhalation of volatile substances, direct skin contact with a dyed cloth or the small children who are orally exposed by sucking and chewing on textile substances. Moreover, the direct and acid dyes are attached very loosely to the fibers which might be easily migrated from the fabric through the skin and saliva. It also affects the reproductive abilities of the exposed organisms.

The occupational exposure to some of the aromatic amines like benzidine, 2-naphthylamine, and 4-aminobiphenyl causes bladder cancer risk. EU restricts the use of aromatic amines. For example, the use of 1, 4-diamino benzene may cause contact dermatitis and blindness. Aromatic amines are easily mobilised by water or sweat, which is absorbed by the skin and mouth. The azo dyes, which are water-soluble are more dangerous. The industrial effluents with azo dyes cause an increasing level of colon cancer. The azo dyes, which are water-soluble, are more dangerous when metabolized by liver enzymes [30 - 33]. Studies on workers exposed to 2-naphthylamine, benzidine, and 4-aminobiphenyl showed an association between human exposure to aromatic amines with an increased risk of urinary bladder cancer. O-toluidine is suggested as a bladder carcinogen. Workers exposed both to o-toluidine and 4, 4'-methylene bis (2-methylaniline) showed a 62-fold increase in bladder cancer risk [34 - 37].

During metabolism, the azo dyes are cleaved by reduction into aromatic amines, which are highly toxic, causing damage to DNA. The bacteria which are present in the human skin are more efficient to metabolize azodyes into aromatic amines, which are easily penetrated into the skin. The bacteria in the intestine may have the ability to cleave the azodyes using the secretion of enzymes such as azoreductase and nitroreductase. The liver in mammals secreted some enzymes that have the ability to cleave the azo dyes [30, 31, 38 - 41].

The reactive dyes cause immune disorders, skin and respiratory problems. The dyes along with additives and bleaching agents, may increase the severity of the disorders to the affected organisms [42, 43]. Moreover, mordant dyeing using chromium salts and/or the other techniques which use chromium as oxidation or fixing agents are responsible for carcinogenic and mutagenic effects [44].

Disperse dyes are highly responsible for contacting allergy. For example, dermatitis is mainly caused by the use of disperse Blue 106, disperse Blue 124 and disperse Yellow 3 [45 - 49]. Mostly the synthetic fabrics such as polyester, acetate and polyamide are stained with disperse dyes. If the disperse dyes are not properly fixed on the fabric, it may easily come out from the fabric and diffused into the skin of the person who wears the garment. EU ecolabel listed out 19 disperse dyes as allergens. Some of the case reports showedpatients who suffered from textile dermatitis were caused by reactive, acidic and basic dyes in clothing [42, 50 - 53].

## ENVIRONMENTAL STANDARDS

The chemicals present in the garments are passed through the skin because clothes are in close contact with the skin. European Union (EU), 2006, has adopted the Registration, Evaluation, Authorization and Restriction of Chemicals (REACH) to guard human fitness and the surroundings against risks by means of chemical substances. EU added regulations to use, manufacturing, marketing of the restricted substances [54]. In 2002, EU banned the azo dyes that could break down to one of the 24 possible carcinogenic products like aromatic amines. Most of the synthetic dyes are either restricted or regulated under REACH or included in the candidate list, mainly related to substances with carcinogenic and mutagenic effects. Moreover, REACH bans the presence of carcinogenic aryl amines in consumer goods.

## TREATMENT METHODS

The reactive azo dyes (Disperse blue 373, Reactive blue 19, Disperse Orange 37) are not easily degradable due to their chemical stability and synthetic nature and not easily treated by the conventional methods such as trickling filter, flocculation and electrodialysis. The dioxins and furans are formed due to the incomplete combustion of these substances during incineration. It is very difficult to remove because of their stability in aerobic conditions and anaerobic biotreatment causes the synthesis of dangerous aromatic amines. Advanced oxidation processes (AOPs) developed the hydroxyl free radicals by using different oxidants, are able to destroy the components that are not easily degraded by using conventional treatment methods. AOP with a combination of ozone, UV, $TiO_2$, fenton, photo-fenton, hydrogen peroxides ($H_2O_2$) and ultrasonic (US) could be used efficiently for the treatment of textile industry effluent. This technology is more useful to degrade the complex structure with maximum oxidation. Hydroxyl radicals, which are the main oxidative energy, are released from the photolytic separation of $H_2O_2$ in water by UV light [59]. The chromospheres in the dyes are efficiently degraded by hydrogen peroxide and Ultraviolet treatment. The sulphonated azo and anthraquinone dyes are completely decolourized by UV irradiation along with $H_2O_2$. Fenton's oxidation by using the $H_2O_2/Fe_2^+$ system is one of the efficient methods for treating textile dyeing industry effluent, which is done by the formation of hydroxyl radicals and followed by ferric coagulation. Adsorption is an effective treatment method for textile wastewater in order to remove the dyes. The adsorbent with a highly porous surface adsorbs the compound to be removed. Activated carbon is a good adsorbent. Biosorption of dyes was studied by using various agricultural wastes (bagasse, ground nutshell and corncobs) and industrial wastes (coal ashes and wood chips), which have the efficiency to absorb the dyes from textile effluents with a removal capacity of 40-90% of basic dyes (40-90%)

and direct dyes (40%). Reverse Osmosis is used to remove hydrolysed reactive dyes. The ozonisation treatment irradiation is more useful to treat the textile effluent. The dyes are coagulated by coagulants such as aluminum and iron slats. The addition of the lime, ferrous/ferric sulphate, ferric chloride, aluminium sulphate/chloride enhanced the precipitation of dyes in textile effluents [55 - 59]. The maximum dyes are removed from effluent by using the electrocoagulation method, which has many advantages, such as the removal of very small colloidal particles. It is a very simple and low-cost process than other methods [60].

## Biological Treatment

The dyes are removed by biosorption/biodegradation using bacteria, fungi, yeast and algae either by single species/mixed consortium/immobilized state. The dead algae have shown maximum efficiency for the absorption of dyes. The biological treatment is ecofriendly and cheapest than other treatments. Table **3** shows the ability of microbes to biosorb/degrade different types of dyes.

Table 3. Microorganisms and their efficiency for biosorption/degradation of dyes.

| Name of Microorganisms | Dye |
|---|---|
| **Bacteria** | |
| *Corynebacterium glutamicum* | Reactive Red 4 [61] |
| *Bacillus weihenstephanensis RI12* | Congo red [62] |
| *Geobacillus stearothermophilus* UCP 986 | Orange II [63] |
| *Bacillus pumilus* HKG212 | Remazol Navy Blue [64] |
| **Bacterial consortium** | |
| *Enterobacter cloacae* and *Enterococcus casseliflavus* | Orange II [65] |
| *Bacillus odysseyi* SUK3, *Proteus* sp. SUK7 and *Morganella morganii* SUK5 | Reactive Blue 59 [66] |
| *Bacillus flexus* strain NBN2 (SY1), *Bacillus cereus* strain AGP-03 (SY2), *Bacillus cytotoxicus* NVH 391-98 (SY3) and *Bacillus* sp. L10 (SY4) | Direct Blue 151 Direct Red 31 [67] |
| *Paenibacillus polymyxa, Micrococcus luteus* and *Micrococcus* sp. | Reactive Violet 5R [68] |
| *Aeromonas caviae, Protues mirabilis* and *Rhodococcus globerulus* | Acid Orange 7 [69] |
| **Fungi** | |
| *Phanerochaete chrysosporium* | Direct Red [70] |
| *Pleurotus eryngii* F032 | Reactive Black 5 [71] |
| *Trametes* sp. | Orange II, Brilliant Blue R 250 [72] |

(Table 3) cont.....

| Name of Microorganisms | Dye |
|---|---|
| *Fusarium oxysporum* | Yellow GAD [73] |
| *Aspergillus niger* | Congo Red [74] |
| **Yeast** | |
| *Candida Oleophila* | Reactive Black 5 [75] |
| *Candida albicans* | Direct Violet 51 [76] |
| *Candida tropicalis* | Violet 3 [77] |
| *Saccharomyces cerevisiae* MTCC463 | Malachite Green [78] |
| **Algae** | |
| *Oscillatoria curviceps* | Acid Black [79] |
| *Chara* and *Scenedesmus obliquus* | Congo Red Crystal Violet [80] |

The enzymes (lignin peroxidase, manganese peroxidase and laccase), which are synthesized by bacteria and fungi, could degrade the dyes efficiently. Some of the bacteria can reduce azo compounds to aromatic amines in the presence of azo reductases, laccase, lignin peroxidase [81 - 83]. The azo dyes such as Basic Red 46, Basic Yellow 19, Acid Red 151 and Basic Blue 41 are degraded efficiently by using an aerobic biofilm system in which the efficiency was further improved by adding activated carbon in an aeration tank. The azo dyes are also degraded by an anaerobic method in which the electrons could reduce the bonds in azo dyes that lead to the decolourization of the effluent with the release of toxic amines [84].

## CONTROL MEASURES

The wastewater should be recycled and reused by the textile industries for various processing. The customers should have awareness of the toxic effects of synthetic dyes and avoid the products based on it. The industries should prefer to use alternative ecofriendly dyes, which are safe for both human beings and the environment. Air dyeing technology is an advanced dyeing method that uses only air to dye the fabrics instead of water. It minimizes the use of more water for processing and reduces the pollution effects on the environment. The chlorine bleaching of fabrics could be replaced by alternative hydrogen peroxide or ozone treatment. For the finishing of fabrics, ecosafe natural material such as beeswax could be used as an alternative to toxic chemicals. During sizing, the toxic polyvinyl alcohol is replaced with CMC (Carboxy Methyl Cellulose). The textile and leather industries which use nearly 42 benzidine dyes are restricted in India from February 1, 1993. Food industries should use natural dyes as an alternative to synthetic dyes. The other industries, such as cosmetics, drug, paint and paper industries, which depend on synthetic dyes, should find ecofriendly alternative to synthetic dyes [85].

Environmental laws should be followed, which regulate the effects of industrial effluents that solve the environmental and health problems linked with synthetic dyes. Natural dyes are both environment and human safe. So the research should be aimed at enhancing the quantity and quality of dyes, which are extracted from the efficient microbes.

## CONSENT FOR PUBLICATION

Not applicable.

## CONFLICT OF INTEREST

The authors confirm that this chapter contents have no conflict of interest.

## ACKNOWLEDGEMENTS

Declared none.

## REFERENCES

[1]     Bafana A, Devi SS, Chakrabarti T. Azo dyes: Past, present and the future. Environ Rev 2011; 19: 350-70.
        [http://dx.doi.org/10.1139/a11-018]

[2]     Reddy SS, Kotaiah B, Reddy NSP. Color pollution control in textile dyeing industry effluents using tannery sludge derived activated carbon. Bull Chem Soc Ethiop 2008; 22(3): 369-78.

[3]     Vassileva V, Valcheva E, Zheleva Z. The kinetic model of reactive dye fixation on cotton fibers. J University of Chemical Technology and Metallurgy 2008; 43(3): 323-6.

[4]     Zollinger H. Color chemistry. Syntheses, properties and applications of organic dyes and pigments. Weinheim: VCH Publishers 2003.

[5]     Moore SB, Ausley LW. Systems thinking and green chemistry in the textile industry: Concepts, technologies and benefits. J Clean Prod 2004; 12: 585-601.
        [http://dx.doi.org/10.1016/S0959-6526(03)00058-1]

[6]     Zhao J, Wu T, Wu K. Kyoko Oikawa, Hisao Hidaka and Nick Serpone. Photoassisted degradation of dye pollutants. 3. degradation of the cationic dye rhodamine b in aqueous anionic surfactant/$TiO_2$ dispersions under visible light irradiation: evidence for the need of substrate adsorption on $TiO_2$ particles. Environ Sci Technol 1998; 32(16): 2394-400.
        [http://dx.doi.org/10.1021/es9707926]

[7]     Shakir K, Elkafrawy AF, Ghoneimy HF, Elrab Beheir SG, Refaat M. Removal of rhodamine B (a basic dye) and thoron (an acidic dye) from dilute aqueous solutions and wastewater simulants by ion flotation. Water Res 2010; 44(5): 1449-61.
        [http://dx.doi.org/10.1016/j.watres.2009.10.029] [PMID: 19942250]

[8]     Chequer FMD, Venâncio VP, Bianchi ML, Antunes LMG. Genotoxic and mutagenic effects of erythrosine B, a xanthene food dye, on HepG2 cells. Food Chem Toxicol 2012; 50(10): 3447-51.
        [http://dx.doi.org/10.1016/j.fct.2012.07.042] [PMID: 22847138]

[9]     Chanlon S, Joly-Pottuz L, Chatelut M, Vittori O, Cretier JL. Determination of carmoisine, allura red and ponceau 4R in sweets and soft drinks by differential pulse polarography. J Food Compos Anal 2005; 18(6): 503-15.

[http://dx.doi.org/10.1016/j.jfca.2004.05.005]

[10]   Puttemans ML, Dryon L, Massart DL. Extraction of organic acids by ion-pair formation with tri--octylamine. Part V. Simultaneous determination of synthetic dyes, benzoic acid, sorbic acid, and saccharin in soft drinks and lemonade syrups. J Assoc Off Anal Chem 1984; 67(5): 880-5.
       [http://dx.doi.org/10.1093/jaoac/67.5.880] [PMID: 6501150]

[11]   Suzuki S, Shirao M, Aizawa M, Nakazawa H, Sasa K, Sasagawa H. Determination of synthetic food dyes by capillary electrophoresis. J Chromatogr A 1994; 680(2): 541-7.
       [http://dx.doi.org/10.1016/0021-9673(94)85153-0] [PMID: 7981834]

[12]   Purba MK, Agrawal N, Shukla SK. Detection of non-permitted food colors in edibles. Journal of Forensic Research 2015; (S4): 3.

[13]   Przystaś W, Zabłocka-Godlewska E, Grabińska-Sota E. Biological removal of azo and triphenylmethane dyes and toxicity of process by-products. Water Air Soil Pollut 2012; 223(4): 1581-92.
       [http://dx.doi.org/10.1007/s11270-011-0966-7] [PMID: 22593606]

[14]   Abdullah SU, Badaruddin M, Sayeed SA, Ali R, Riaz MN. Binding ability of Allura Red with food proteins and its impact on protein digestibility. Food Chem 2008; 110(3): 605-10.
       [http://dx.doi.org/10.1016/j.foodchem.2008.02.049]

[15]   Millichap JG, Yee MM. The diet factor in attention-deficit/hyperactivity disorder. Pediatrics 2012; 129(2): 33-330.

[16]   Kanarek RB. Artificial food dyes and attention deficit hyperactivity disorder. Nutr Rev 2011; 69(7): 385-91.
       [http://dx.doi.org/10.1111/j.1753-4887.2011.00385.x] [PMID: 21729092]

[17]   Stevens LJ, Kuczek T, Burgess JR, Hurt E, Arnold LE. Dietary sensitivities and ADHD symptoms: thirty-five years of research. Clin Pediatr (Phila) 2011; 50(4): 279-93.
       [http://dx.doi.org/10.1177/0009922810384728] [PMID: 21127082]

[18]   Brown MA, De Vito SC. Predicting azo dye toxicity. Crit Rev Environ Sci Technol 1993; 23(3): 249-324.
       [http://dx.doi.org/10.1080/10643389309388453]

[19]   Alim N, Zahra N, Akhlaq F. Detection of Sudan dyes in different spices. Pak J Food Sci 2015; 25(3): 144-9.

[20]   Brown JP, Roehm GW, Brown RJ. Mutagenicity testing of certified food colors and related azo, xanthene and triphenylmethane dyes with the Salmonella/microsome system. Mutat Res 1978; 56(3): 249-71.
       [http://dx.doi.org/10.1016/0027-5107(78)90192-6] [PMID: 342943]

[21]   Chung KT, Cerniglia CE. Mutagenicity of azo dyes: Structure-activity relationships. Mutation Research/Reviews in Genetic Toxicology 1992; 277(3): 201-20.

[22]   Pinheiro HM, Touraud E, Thomas O. Aromatic amines from azo dye reduction: status review with emphasis on direct UV spectrophotometric detection in textile industry wastewaters. Dyes Pigments 2004; 61(2): 121-39.
       [http://dx.doi.org/10.1016/j.dyepig.2003.10.009]

[23]   Chen H. Recent advances in azo dye degrading enzyme research. Curr Protein Pept Sci 2006; 7(2): 101-11.
       [http://dx.doi.org/10.2174/138920306776359786] [PMID: 16611136]

[24]   Puvaneswari N, Muthukrishnan J, Gunasekaran P. Toxicity assessment and microbial degradation of azo dyes. Indian J Exp Biol 2006; 44(8): 618-26.
       [PMID: 16924831]

[25]   Luo H, Ning XA, Liang X, Feng Y, Liu J. Effects of sawdust-CPAM on textile dyeing sludge

dewaterability and filter cake properties. Bioresour Technol 2013; 139: 330-6.
[http://dx.doi.org/10.1016/j.biortech.2013.04.035] [PMID: 23665695]

[26]    Tsuboy MS, Angeli JP, Mantovani MS, Knasmüller S, Umbuzeiro GA, Ribeiro LR. Genotoxic, mutagenic and cytotoxic effects of the commercial dye CI Disperse Blue 291 in the human hepatic cell line HepG2. Toxicol *In Vitro.* 2007; 21(8): 1650-5.
[http://dx.doi.org/10.1016/j.tiv.2007.06.020] [PMID: 17728095]

[27]    Prasad AS, Rao KV. Aerobic biodegradation of Azo dye by *Bacillus cohnii* MTCC 3616; an obligately alkaliphilic bacterium and toxicity evaluation of metabolites by different bioassay systems. Appl Microbiol Biotechnol 2013; 97(16): 7469-81.
[http://dx.doi.org/10.1007/s00253-012-4492-3] [PMID: 23070653]

[28]    Caritá R, Marin-Morales MA. Induction of chromosome aberrations in the *Allium cepa* test system caused by the exposure of seeds to industrial effluents contaminated with azo dyes. Chemosphere 2008; 72(5): 722-5.
[http://dx.doi.org/10.1016/j.chemosphere.2008.03.056] [PMID: 18495201]

[29]    Seesuriyachan P, Takenaka S, Kuntiya A, Klayraung S, Murakami S, Aoki K. Metabolism of azo dyes by *Lactobacillus casei* TISTR 1500 and effects of various factors on decolorization. Water Res 2007; 41(5): 985-92.
[http://dx.doi.org/10.1016/j.watres.2006.12.001] [PMID: 17254626]

[30]    Ben Mansour H, Corroler D, Barillier D, Ghedira K, Chekir L, Mosrati R. Evaluation of genotoxicity and pro-oxidant effect of the azo dyes: acids yellow 17, violet 7 and orange 52, and of their degradation products by *Pseudomonas putida* mt-2. Food Chem Toxicol 2007; 45(9): 1670-7.
[http://dx.doi.org/10.1016/j.fct.2007.02.033] [PMID: 17434654]

[31]    Forgacs E, Cserháti T, Oros G. Removal of synthetic dyes from wastewaters: a review. Environ Int 2004; 30(7): 953-71.
[http://dx.doi.org/10.1016/j.envint.2004.02.001] [PMID: 15196844]

[32]    Lima RO Alves de, Bazo AP, Salvadori DMF, Rech CM, *et al.* Mutagenic and carcinogenic potential of a textile azo dye processing plant effluent that impacts a drinking water source. Mutat Res 2007; 626(1-2): 53-60.

[33]    Platzek T, Lang C, Grohmann G, US GI, W. Baltes. Formation of a carcinogenic aromatic mine from an azo dye by human skin bacteria *in Vitro.* Hum Exp Toxicol 1999; 18(9): 552-9.

[34]    Glenn Talaska. Aromatic amines and human urinary bladder cancer: exposure sources and epidemiology. J Environmental Science and Health, Part C, 2003; 21(1): 29-43.

[35]    Rubino GF, Scansetti G, Piolatto G, Pira E. The carcinogenic effect of aromatic amines: an epidemiological study on the role of o-toluidine and 4,4'-methylene bis (2-methylaniline) in inducing bladder cancer in man. Environ Res 1982; 27(2): 241-54.
[http://dx.doi.org/10.1016/0013-9351(82)90079-2] [PMID: 7084156]

[36]    Carliell CM, Barclay SJ, Shaw C, Wheatley AD, Buckley CA. The effect of salts used in textile dyeing on microbial decolourisation of a reactive azo dye. Environ Technol 1998; 19(11): 1133-7.
[http://dx.doi.org/10.1080/09593331908616772]

[37]    Combes RD, Haveland-Smith RB. A review of the genotoxicity of food, drug and cosmetic colours and other azo, triphenylmethane and xanthene dyes. Mutat Res 1982; 98(2): 101-248.
[http://dx.doi.org/10.1016/0165-1110(82)90015-X] [PMID: 7043261]

[38]    de Aragão Umbuzeiro G, Freeman H, Warren SH, Kummrow F, Claxton LD. Mutagenicity evaluation of the commercial product CI Disperse Blue 291 using different protocols of the Salmonella assay. Food Chem Toxicol 2005; 43(1): 49-56.
[http://dx.doi.org/10.1016/j.fct.2004.08.011] [PMID: 15582195]

[39]    Arlt VM, Glatt H, Muckel E, *et al.* Metabolic activation of the environmental contaminant 3-nitrobenzanthrone by human acetyltransferases and sulfotransferase. Carcinogenesis 2002; 23(11):

1937-45.
[http://dx.doi.org/10.1093/carcin/23.11.1937] [PMID: 12419844]

[40] Lisi P, Stingeni L, Cristaudo A. Clinical and epidemiological features of textile contact dermatitis: An Italian multicentre study. Contact Dermatitis 2014; 70(6): 344-50.

[41] Ale IS, Maibacht HA. Diagnostic approach in allergic and irritant contact dermatitis. Expert Rev Clin Immunol 2010; 6(2): 291-310.
[http://dx.doi.org/10.1586/eci.10.4] [PMID: 20402391]

[42] Zeng M, Xiao F, Zhong X, *et al.* Reactive oxygen species play a central role in hexavalent chromium-induced apoptosis in Hep3B cells without the functional roles of p53 and caspase-3. Cell Physiol Biochem 2013; 32(2): 279-90.
[http://dx.doi.org/10.1159/000354436] [PMID: 23942225]

[43] Ryberg K, Goossens A, Isaksson M, *et al.* Is contact allergy to disperse dyes and related substances associated with textile dermatitis? Br J Dermatol 2009; 160(1): 107-15.
[http://dx.doi.org/10.1111/j.1365-2133.2008.08953.x] [PMID: 19067698]

[44] Malinauskiene L, Bruze M, Ryberg K, Zimerson E, Isaksson M. Contact allergy from disperse dyes in textiles: a review. Contact Dermat 2013; 68(2): 65-75.
[http://dx.doi.org/10.1111/cod.12001] [PMID: 23289879]

[45] Caliskaner Z, Kartal O, Baysan A, *et al.* A case of textile dermatitis due to disperse blue on the surgical wound. Hum Exp Toxicol 2012; 31(1): 101-3.
[http://dx.doi.org/10.1177/0960327111424300] [PMID: 22027509]

[46] Narganes LM, Sambucety PS, Gonzalez IR, Rivas MO, Prieto MA. Lymphomatoid dermatitis caused by contact with textile dyes. Contact Dermat 2013; 68(1): 62-4.
[http://dx.doi.org/10.1111/j.1600-0536.2012.02164.x] [PMID: 23227872]

[47] Hession MT, Scheinman PL. Lymphomatoid allergic contact dermatitis mimicking cutaneous T cell lymphoma. Dermatitis 2010; 21(4): 220.
[http://dx.doi.org/10.2310/6620.2010.10010] [PMID: 20646676]

[48] Wentworth AB, Richardson DM, Davis MD. Patch testing with textile allergens: the mayo clinic experience. Dermatitis 2012; 23(6): 269-74.
[http://dx.doi.org/10.1097/DER.0b013e318277ca3d] [PMID: 23169208]

[49] Slodownik D, Williams J, Tate B, *et al.* Textile allergy--the Melbourne experience. Contact Dermat 2011; 65(1): 38-42.
[http://dx.doi.org/10.1111/j.1600-0536.2010.01861.x] [PMID: 21309788]

[50] Moreau L, Goossens A. Allergic contact dermatitis associated with reactive dyes in a dark garment: a case report. Contact Dermat 2005; 53(3): 150-4.
[http://dx.doi.org/10.1111/j.0105-1873.2005.00663.x] [PMID: 16128754]

[51] Curr N, Nixon R. Allergic contact dermatitis to basic red 46 occurring in an HIV-positive patient. Australas J Dermatol 2006; 47(3): 195-7.
[http://dx.doi.org/10.1111/j.1440-0960.2006.00272.x] [PMID: 16867003]

[52] Regulation (EC) No 1907/2006 of the European Parliament and of the Council of 18 December 2006 concerning the Registration, Evaluation, Authorisation and Restriction of Chemicals (REACH), establishing a European Chemicals Agency.

[53] Leme DM, de Oliveira GA, Meireles G, dos Santos TC, Zanoni MV, de Oliveira DP. Genotoxicological assessment of two reactive dyes extracted from cotton fibres using artificial sweat. Toxicol *In Vitro*. 2014; 28(1): 31-8.
[http://dx.doi.org/10.1016/j.tiv.2013.06.005] [PMID: 23811265]

[54] Carneiro PA, Umbuzeiro GA, Oliveira DP, Zanoni MVB. Assessment of water contamination caused by a mutagenic textile effluent/dyehouse effluent bearing disperse dyes. J Hazard Mater 2010; 174(1-3): 694-9.

[http://dx.doi.org/10.1016/j.jhazmat.2009.09.106] [PMID: 19853375]

[55]   Arun Prasad AS, Bhaskara Rao KV. Physico chemical characterization of textile effluent and screening for dye decolorizing bactéria. Global J Biotechnology and Biochemistry 2010; 5(2): 80-6.

[56]   Robinson T, McMullan G, Marchant R, Nigam P. Remediation of dyes in textile effluent: a critical review on current treatment technologies with a proposed alternative. Bioresour Technol 2001; 77(3): 247-55.
[http://dx.doi.org/10.1016/S0960-8524(00)00080-8] [PMID: 11272011]

[57]   Hao OJ, Kim H, Chiang PC. Decolorization of wastewater. Crit Rev Environ Sci Technol 2000; 30(4): 449-505.
[http://dx.doi.org/10.1080/10643380091184237]

[58]   dos Santos AB, Cervantes FJ, van Lier JB. Review paper on current technologies for decolourisation of textile wastewaters: perspectives for anaerobic biotechnology. Bioresour Technol 2007; 98(12): 2369-85.
[http://dx.doi.org/10.1016/j.biortech.2006.11.013] [PMID: 17204423]

[59]   Firmino PIM, da Silva ME, Cervantes FJ, dos Santos AB. Colour removal of dyes from synthetic and real textile wastewaters in one- and two-stage anaerobic systems. Bioresour Technol 2010; 101(20): 7773-9.
[http://dx.doi.org/10.1016/j.biortech.2010.05.050] [PMID: 20542688]

[60]   Mollah MY, Morkovsky P, Gomes JA, Kesmez M, Parga J, Cocke DL. Fundamentals, present and future perspectives of electrocoagulation. J Hazard Mater 2004; 114(1-3): 199-210.
[http://dx.doi.org/10.1016/j.jhazmat.2004.08.009] [PMID: 15511592]

[61]   Won SW, Choi SB, Yun YS. Interaction between protonated waste biomass of *Corynebacterium glutamicum* and anionic dye Reactive Red 4. Col Surf A 2005; 262: 175-280.
[http://dx.doi.org/10.1016/j.colsurfa.2005.04.028]

[62]   Mnif I, Fendri R, Ghribi D. Biosorption of Congo Red from aqueous solution by *Bacillus weihenstephanensis* RI12; effect of SPB1 biosurfactant addition on biodecolorization potency. Water Sci Technol 2015; 72(6): 865-74.
[http://dx.doi.org/10.2166/wst.2015.288] [PMID: 26360745]

[63]   Evangelista-Barreto NS, Albuquerque CD, Vieira RHSF, Campos-Takaki GM. Co-metabolic decolorizationof the reactive azo dye Orange II by *Geobacillus stearothermophilus* UCP 986. Text Res J 2009; 79: 1266-73.
[http://dx.doi.org/10.1177/0040517508087858]

[64]   Das A, Mishra S, Verma VK. Enhanced biodecolorization of textile dye remazol navy blue using an isolated bacterial strain *Bacillus pumilus* HKG212 under improved culture conditions. J Biochem Technol 2015; 6: 962-9.

[65]   Chan GF, Rashid NAA, Koay LL, Chang SY, Tan WL. Identification and optimization of novel NAR-1 bacterial consortium for the biodegradation of Orange II. Insight Biotechnol 2011; 1: 7-16.
[http://dx.doi.org/10.5567/IBIOT-IK.2011.7.16]

[66]   Patil PS, Shedbalkar UU, Kalyani DC, Jadhav JP. Biodegradation of Reactive Blue 59 by isolated bacterial consortium PMB11. J Ind Microbiol Biotechnol 2008; 35(10): 1181-90.
[http://dx.doi.org/10.1007/s10295-008-0398-6] [PMID: 18661161]

[67]   Lalnunhlimi S, Krishnaswamy V. Decolorization of azo dyes (Direct Blue 151 and Direct Red 31) by moderately alkaliphilic bacterial consortium. Braz J Microbiol 2016; 47(1): 39-46.
[http://dx.doi.org/10.1016/j.bjm.2015.11.013] [PMID: 26887225]

[68]   Moosvi S, Kher X, Datta M. Isolation, characterization and decolorization of textile dyes by a mixed bacterial consortium JW-2. Dyes Pigments 2007; 74: 723-9.
[http://dx.doi.org/10.1016/j.dyepig.2006.05.005]

[69]   Joshi T, Iyengar L, Singh K, Garg S. Isolation, identification and application of novel bacterial

consortium TJ-1 for the decolourization of structurally different azo dyes. Bioresour Technol 2008; 99(15): 7115-21.
[http://dx.doi.org/10.1016/j.biortech.2007.12.074] [PMID: 18289845]

[70]  Sen K, Pakshirajan K, Santra SB. Modeling the biomass growth and enzyme secretion by the white rot fungus *Phanerochaete chrysosporium*: a stochastic-based approach. Appl Biochem Biotechnol 2012; 167(4): 705-13.
[http://dx.doi.org/10.1007/s12010-012-9720-x] [PMID: 22588736]

[71]  Hadibarata T, Adnan LA, Yusoff ARM, Yuniarto A. Rubiyanto, Zubir MMFA, Khudhair AB, Teh ZC and Naser MA. Microbial decolorization of an azo dye Reactive Black 5 using white rot fungus *Pleurotus eryngii* F032. Water Air Soil Pollut 2013; 224: 1595-604.
[http://dx.doi.org/10.1007/s11270-013-1595-0]

[72]  Grinhut T, Salame TM, Chen Y, Hadar Y. Involvement of ligninolytic enzymes and Fenton-like reaction in humic acid degradation by *Trametes* sp. Appl Microbiol Biotechnol 2011; 91(4): 1131-40.
[http://dx.doi.org/10.1007/s00253-011-3300-9] [PMID: 21541787]

[73]  Porri A, Baroncelli R, Guglielminetti L, *et al. Fusarium oxysporum* degradation and detoxification of a new textile-glycoconjugate azo dye (GAD). Fungal Biol 2011; 115(1): 30-7.
[http://dx.doi.org/10.1016/j.funbio.2010.10.001] [PMID: 21215952]

[74]  Karthikeyan K, Nanthakumar K, Shanthi K, Lakshmanaperumalsamy P. Response surface methodology for optimization of culture conditions for dye decolorization by a fungus, *Aspergillus niger* HM11 isolated from dye affected soil. Iran J Microbiol 2010; 2(4): 213-22.
[PMID: 22347575]

[75]  Lucas MS, Amaral C, Sampaio A, Peres JA, Dias AA. Biodegradation of diazo dye Reactive Black 5 by a wild isolate of *Candida oleophila.* Enzyme Microb Technol 2006; 39: 51-5.
[http://dx.doi.org/10.1016/j.enzmictec.2005.09.004]

[76]  Vitor V, Corso CR. Decolorization of textile dye by isolated *Candida albicans* from industrial effluents. J Ind Microbiol Biotechnol 2008; 35: 1353-7.
[http://dx.doi.org/10.1007/s10295-008-0435-5] [PMID: 18712543]

[77]  Charumathi D, Das N. Bioaccumulation of synthetic dyes by *Candida tropicalis* growing in sugarcane bagasse extract medium. Adv Biol Res (Faisalabad) 2010; 4: 233-40.

[78]  Jadhav JP, Govindwar SP. Biotransformation of malachite green by *Saccharomyces cerevisiae* MTCC 463. Yeast 2006; 23(4): 315-23.
[http://dx.doi.org/10.1002/yea.1356] [PMID: 16544273]

[79]  Priya B, Uma L, Ahamed AK, Subramanian G, Prabaharan D. Ability to use the diazo dye, C.I. Acid Black 1 as a nitrogen source by the marine cyanobacterium Oscillatoria curviceps BDU92191. Bioresour Technol 2011; 102(14): 7218-23.
[http://dx.doi.org/10.1016/j.biortech.2011.02.117] [PMID: 21571528]

[80]  Patil KJ, Mahajan RT, Lautre HK, Hadda TB. Bioprecipitation and biodegradation of fabric dyes by using *Chara* sp. and *Scenedesmus obliquus.* J Chem Pharm Res 2015; 7: 783-91.

[81]  Stolz A. Basic and applied aspects in the microbial degradation of azo dyes. ApplMicrobiolBiotechnol 2001; 56: 69-80.

[82]  Misal SA, Lingojwar DP, Shinde RM, Gawai KR. Purification and characterization of azoreductase from alkaliphilic strain *Bacillus badius.* Process Biochem 2011; 46: 1264-9.
[http://dx.doi.org/10.1016/j.procbio.2011.02.013]

[83]  Hai FI, Yamamoto K, Nakajima F, Fukushi K. Application of a GAC-coated hollow fiber module to couple enzymatic degradation of dye on membrane to whole cell biodegradation within a membrane bioreactor. J Membr Sci 2012; 389: 67-75.
[http://dx.doi.org/10.1016/j.memsci.2011.10.016]

[84]  Banat ME, Nigam P, Singh D, Marchant R. Microbial decolourization of textile dye containing

effluents, a review. Bioresour Technol 1996; 58: 217-27.
[http://dx.doi.org/10.1016/S0960-8524(96)00113-7]

[85]    Rita Kant. Textile dyeing industry an environmental hazard. Natural Science 2012; 4(1): 22-6.
[PMID: 22107826]

# CHAPTER 2

# Microplastics Pollution

**B. Sathya Priya[1,*] and T. Stalin[2]**

[1] *Department of Environmental Sciences, Bharathiar University, Coimbatore - 641 046, Tamil Nadu, India*

[2] *Forestry Research and Development Unit, Molecular Biology Division, Karur-639 136, Tamil Nadu, India*

**Abstract:** The marine ecosystem is highly contaminated by plastic debris, which is the major part of marine debris. Plastics are nondegradable and persistent in the environment for many centuries. The products which we are using in our day to day life have microplastics and the dumping of plastic waste is converted into microplastics by weathering and degradation. The microplastics' contamination in the ocean and other products is an emerging global concern that has severe impacts on the environment and health problems to the biota. The ingestion of microplastics by the marine biota and the human being's also by consuming seafood and other products leads to severe health disorders. Moreover, the microplastics have a special feature of adsorbing the toxicants in the environment and act as a vehicle for the transfer of those toxicants to which the organisms are exposed. The top predators may be subjected to greater risk due to the ingestion of microplastics. This study deals with the sources of microplastics, their impacts on marine biota and human beings, the management and control measures of microplastic pollution in the environment.

**Keywords:** Eco-safe, Marine Biota, Marine Debris, Mesoplastics, Microliter, Microplastics, POPs.

## INTRODUCTION

Plastics play an important role in our everyday life and are widely used for different applications due to their lightweight, strong, versatile, easy handling, bioinertness, less cost and moisture barrier nature. However, plastics are not easily degradable and become hazardous to the environment. The population explosion increases the usage and dumping of plastic materials that cause a serious threat to the marine ecosystem [1]. Moreover, the material of choice for plastics increases the number of package resins. Plastic litter is found in each part of the marine environment: coastal sea surface, throughout the water section [2].

* **Corresponding author B. Sathya Priya:** Department of Environmental Sciences, Bharathiar University, Coimbatore - 641 046, Tamil Nadu, India; Tel: 8825688512; E-mail: sbspriya11@gmail.com

**J. Senthil Kumar, P. Ponmurugan & A. Vinothkanna (Eds.)**

Plastic litter is increasing every year all over the world. The plastics which are dumped continuously and which exist already in the marine environment increase the debris [3].

Microplastics or microlitter are the minute part of the plastics, which have a diameter of less than 5 mm and are not visible to the naked eye. It is a man-made litter that is present in the oceans for many decades [4, 5]. The particles, which are larger such as virgin resin pellets, are called mesoplastics, and are originated from large plastic debris [6, 7]. The plastic particle which has a dm of greater than 25 mm is macroplastic, 5 - 25 mm is mesoplastic, less than 5 mm is microplastic and less than 1mm is termed as small microplastic, which seriously affects the vertebrates and invertebrates [8]. Plastics are ingested by marine organisms, either directly or indirectly with food [9]. The microplastics are ingested, reducing the reproductive activity, blockage of the intestines and leading to death. Some of the plastic debris is consumed from land-based sources and remaining from marine, especially fishing industries. Plastic gears are mostly used in fisheries. Microplastics are easily contaminated by toxic POPs (Persistent Organic Pollutants) and enter into the marine food chain by biomagnification [10]. Beach plastic debris consists of polystyrene, polypropylene, polyvinyl chloride and polyethylene [11].

Microplastics are the important vehicles for carrying toxic chemicals such as bisphenol A (BPA), polychlorinated biphenyls (PCB) and polycyclic aromatic hydrocarbons (PAHs) that are persistent and cause a threat to the environment [12]. The chemicals used in the manufacturing of plastics are highly toxic and cause adverse effects such as neurotoxicity, disrupting the endocrine system and cancer to human beings and other organisms [13 - 15]. The ingestion of microplastics into the marine biota is followed by the transfer of toxic pollutants and additives to the tissues and causes disorders in different organs. The microplastics are adsorbed to membranes and alter the functions of the cells [16].

## ORIGIN OF MICROPLASTICS

In general, the microplastics are originated from the household, industrial and plastic materials released from the cruise ships. During runoff, microplastics that are present in various products are transported with sewage and enter into the marine habitats. During weathering, the macroplastics are broken down into microplastics and a large quantity of minute polyethylene and polystyrene substances are accumulated in the marine region.

### Primary Microplastics

They are minute in size manufactured directly and used in different fields such as detergents, cleaning products, and cosmetics. They are also used as additives in glues, coating materials, preservative agents for fruits, lubricating agents, pigments carrier, and water softening and pharmaceutical products. They also originate from the ship breaking industry.

## Secondary Microplastics

These are generated from already existing larger plastic debris due to weathering and degradation. Comparatively, the seawater and sediment beaches have less microplastic contents [1].

## DEGRADATION OF PLASTICS

The weight of the plastic polymer is reduced due to degradation and it makes the plastics brittle and either minute fragments or powdered minute particles. The microbes react with the carbon content and further convert the fragments into $CO_2$, which is termed as complete mineralization [17].

## Types of Degradation

  i. **Photodegradation**: The polymer in the plastic is degraded with exposure to sunlight.
  ii. **Biodegradation**: The microorganisms such as bacteria and fungi can degrade the plastic by their metabolic and enzymatic activities.
  iii. **Thermal degradation**: The high temperature is responsible for the degradation of plastics.
  iv. **Thermooxidative degradation**: The moderate temperature enhanced the slow oxidative breakdown of polymers.
  v. **Hydrolysis**: The plastics are degraded when they are reacted with water. This process is very rare and slow.

The ultraviolet-B radiation of sunlight initiated photo-oxidative degradation, which is continued without the need for further radiation. The plastic material such as fishing gear with light stabilizer is not easily degraded by sunlight. The oxidation and hydrolysis process occurs very slowly. It takes a longer time for the degradation of plastics in the marine environment and becomes a global threat. The rate of biodegradation of plastics is high on beaches than in sediments. Similarly, the plastics floated on the water have very slow degradation due to less temperature and reduced oxygen levels [5]. Microplastics have more surface area than macroplastics; hence more contaminants may be absorbed on its surface. They have more surface area, which absorbs and holds an enormous quantity of the toxicants such as POPs (Persistent Organic Pollutants), including PCBs (Poly Chlorinated Biphenyls) and PAHs (Polycyclic Aromatic Hydrocarbons). They act as a carrier of toxic pollutants and transport easily due to its buoyancy from the lowest to higher trophic levels [7, 10, 17].

## FACTORS AFFECTING MICROPLASTICS INGESTION

### Abundance

The improper disposal and dumping of waste in the marine region increase the abundance of microplastics. The easy availability of microplastics is either from primary sources such as industrial, pharmaceutical, cosmetics and household products or by weathering the already dumped macroplastics degraded into microplastics [14]. The high level of microplastics content may increase the chance of ingestion by the marine biota and cause severe effects.

### Colour

The microplastics with colors (white, yellow, *etc*.) are selected by the marine biota, such as fishes, due to their resemblance to their prey [18]. For example, some fishes in the Niantic Bay area, New England, ingest only white polystyrene particles by selection. The predators, such as invertebrates in the pelagic region, ingest the microplastics due to their resemblance to food [19].

### Size

It is an important factor that makes the easy availability of microplastics for biota [7]. Some organisms ingest the microplastics along with food directly or indirectly. During normal feeding, the planktivores ingest microplastics or by mistake, they select the particles as their natural prey. The higher organism, such as Mediterranean fin whale *Balaenoptera physalus*, ingests microplastics either directly or indirectly and has a potential risk for the microplastics effects [20].

### Density

The density of the plastic particles will determine the existence of microplastics in the water column. The microplastics which have low density, such as polyethylene, may float on the surface and the high-density particles such as polyvinyl chloride may sink to the sea sediment. Hence both the surface and bottom dwellers are affected and have more potential risk for the ingestion of microplastics. The filter feeders and planktivores, which are available on the surface of the water, have easy contact with the buoyant particles that are with low density.

## Biofouling

It is also an important parameter for making the availability of microplastics to the biota. For example, the biofilm may be formed on the low-density polyethylene Bags, if they exist in water for one week to three weeks. Later on, because of increasing density, they may sink to the bottom. The biofouling rate of plastics depends on the water conditions, the hardness of polymer, texture, and surface energy [21]. The foraging organisms that defoul the plastics may again make the availability to the surface in which this cycle enhances the bioavailability of microplastics from the surface to the bottom of the seafloor [22].

## Food Chain

Microplastics and their toxicity are transferred from lower trophic levels to higher trophic levels through the food chain. The microplastics are observed in Hooker's sea lions and scat of fur seals. It may be due to the trophic link myctophid fish *Electrona subaspera* [23, 24]. The Antarctic fur seals have microplastic particles. The scats of *Arctocephalus tropicalis* and *A. gazella* from Macquarie Island, Australia, evidenced the presence of microplastics. Some fishes found in the coastal water of England have microplastics similar to planktons. The whale species might have the chance of indirectly consuming microplastics either from the water column or through planktonic prey [20]. The microplastics with many additives and the potential of carrying POPs transferred through the food chain or food web cause the biomagnification problem in different trophic levels [25].

## IMPACTS OF MICROPLASTICS ON MARINE ECOSYSTEM

Plastics are a good substratum for the colonization of different marine organisms. The plastic particles which are floating carry invasive and alien species that cause detrimental effects to the marine ecosystem. One side of the plastic is exposed to the sun and another side is occupied with fouling organisms that threaten marine biodiversity. The majority of the marine litter sinks to the bottom of the ocean [1, 7, 9]. It slows down the gaseous exchange, sequestration and affects the carbon cycle in water. It disturbs the habitat on the seafloor. It transfers the accumulated contaminants to the marine organisms for the long term and is a global concern [10]. It also reduces the aesthetic value of the system, habitat of marine organisms, water quality, revenue from fishing and tourism in the ecosystem [3, 26].

During the manufacturing process, the plastics are enriched with many additives such as antioxidant, antimicrobial agent, and light stabilizing agent and flame retardant to increase the quality and long life. The slow degradation of plastics and their resistance to biodegradation causes long term effects on marine organisms [6]. Ingestion is the main exposure route of microplastics for most of the marine species. Ingestion of microplastics occurs either with prey, water filtration and feeding activity of the organisms. Mostly the microplastics and their additives may be transferred from the prey to the predators. The animals that are exposed to microplastics may incorporate them through the gills and digestive tract. Sometimes the microplastics adhere directly to organisms [27 - 29]. In the ocean, the spherical forms of microplastics are common than fibers, fragments, films, and pellet forms.

## IMPACTS OF MICROPLASTICS ON MARINE BIOTA

Plastic litter, including microplastics such as polystyrene particles, causes severe impacts on marine organisms such as fishes, sea turtles, sea birds and mammals [30, 31]. The ingestion of plastics reduced the secretion of gastric enzymes, depleted the feeding stimulus, reduced the uptake of food, lowered steroid hormone levels, prevented stomach contraction, slowing down the movement of food into the small intestine and ended with internal injury or death [32]. The studies also showed that both polyvinyl chloride and polystyrene particles moved from the gut cavity to their lymph and circulatory systems [31]. It causes starvation, suffocation, drowning, reproductive disorders and death of the marine organisms.

The polystyrene, polyethylene microplastics are ingested by pelagic and benthic marine organisms such as small crustaceans, large crustaceans (*Neomysis integer*), copepods (*Acartia* spp.; *Eurytemora affinis*), crab (*Uca rapax*), small and large fishes [33 - 35]. The fish and cetaceans consumed microplastics, which resemble phytoplanktons. It causes internal injuries, intestinal blockage to retard the growth of marine biota. Entanglement by fishing nets may cause drowning and reducing the efficiency of feeding. The fishes also have plastics in their gastrointestinal tracts. The different colored plastic contents resemble the planktons, which are the main food of fish [27, 28]. The plastics with contaminants and toxic additives are transferred through the food chain to the predators [13].

Mostly cetaceans have the chance of ingesting plastic debris along with food. The fishing gear was found out from the gastrointestinal tract of sperm whales (*Physeter macrocephalus*). It blocks the intestine and the abrasions of the plastic debris in the stomach cause rupture in the stomach [36]. The seabirds *(Procellariiformes)*, by mistake, consume plastic caps as their food. The birds could not distinguish plastics and planktons. Hence they consume more plastics mistakenly. Juvenile albatrosses have the chance to ingest more plastic content than adults [37]. It affects the digestive system of birds, reduces the growth and feeding activities. The plastic debris which is deposited in the gut of seabirds and fish may abrade the plastic into microplastics in their gut by muscular contractions and cause blockage in the intestinal region. The microplastics cause damage to vascular tissues and cause cardiac disorders [31, 38].

The plastic debris is highly responsible for the increased number of deaths in sea turtles in which nets, rope plastic bags are observed in the digestive tracts of turtles. The plastic bags are ingested by turtles because they resemble jellyfish as their main food. The leatherback sea turtles (*Dermochelys coriacea*) have been declined for the past two decades due to the ingestion of plastic debris and are noticed on the endangered list. The plastic debris may block the passage of food and female eggs. Most of the sea turtle varieties such as loggerheads (*Caretta caretta)* and green turtles (*Chelonia mydas*) have plastics in their intestinal tracts [39, 40].

The microplastics could be transferred in trophic levels through different food webs [33]. It documented the biomagnification effects of microplastics to the predators at the top level. The laboratory experiments have shown the adverse effects of microplastics on marine organisms such as reduced metabolic and feeding rate, mortality, reduction in predation and swimming activity, neurotoxicity and oxidative stress, damage in the intestine, abnormalities in larva and reduction in fertilization [41 - 43]. Hence the abundance of microplastics in an area severely affects the varieties of different species.

## IMPACTS OF MICROPLASTICS ON HUMAN BEINGS

### Seafood Contaminated With Plastics

Human beings have the chance of consuming microplastics indirectly along with sea or aquaculture food, including crustaceans (brown shrimp), bivalves (oysters, mussels) and different varieties of fishes [28, 44 - 47]. The seafood or aquaculture to nearby marine ecosystems might be contaminated with more plastic debris [48]. The feed used in aquaculture might have microplastics [49]. The plastic contents were found out from the fishes that were kept in the market [50, 51]. Human beings might have the chance of ingesting microplastics indirectly by consuming the seafood such as shellfish, shrimp, bivalves, gastropods, mussels, crustaceans, and fish varieties are accumulated with microplastic contents [43, 45, 50 - 53, 55].

### Other Food Products With Plastic Contamination

Microplastics are the xenobiotics that are persistent in different ecosystems [54]. Moreover, the presence of microplastics in food items is a serious threat to human beings. Hence microplastics are an important emerging issue of great concern.

The European Commission's portal of Rapid Alert System for Food and Feed (RASFF) and European Food Safety Authority's (EFSA) website reported about the presence of microplastics in a variety of food items consumed by human beings [55, 56]. For example, the presence of microplastics or their fragments are observed in the following food products such as honey, sugar, salt, beer, canned sardines and sprats, drinking water, tap water and beverage cartons [57 - 61].

## Implications for Human Food Safety

Microplastics are the important vehicles for carrying toxic chemicals such as bisphenol A (BPA), polychlorinated biphenyls (PCB) and polycyclic aromatic hydrocarbons (PAHs) that are persistent and cause a threat to the environment [13, 15, 42, 50, 62, 63]. The chemicals used in the manufacturing of plastics are highly toxic and cause adverse effects such as neurotoxicity, disrupting the endocrine system and cancer to human beings and other organisms [14, 15]. The ingestion of microplastics into the marine biota is followed by the transfer of toxic pollutants and additives to the tissues and causes disorders in different organs. The microplastics are adsorbed to membranes and alter the functions of the cells [16].

The higher toxicity of phthalates and bisphenolA is evidenced in animal studies [64]. The ability of microplastics to adsorb very toxic metals is observed [62, 65, 71]. Mercury is highly toxic to animals, human beings and its organic form methyl mercury causes biomagnification problems through the food web [66]. Plastisphere represents the presence of microbes and other organisms on plastic debris [67]. It is responsible for the spread of pathogens and exotic species such as *E. coli, Vibrio* sp*., Bacillus cereus, Aeromonas salmonicida* [68, 69]. Hence the organisms on plastic debris may increase the spreading of diseases to animals, human beings, loss of biodiversity and affect the ecosystem and reduce the economic value in that area [67, 68]. Microplastics increase the contamination of seafood [12, 64]. The *in vitro* studies using cerebral and epithelial human cells evidenced that micro and nano plastics may cause cytotoxic effects due to oxidative stress at a cellular level [70].

## Management Methods of Plastic Pollution

Plastics account for the major part of the marine debris. So it is a need to protect the organisms and environment from the effects of microplastics. The plastics that are circulated in the ocean gyres are cleaned by using Ocean Clean-up Array [71]. The plastic contents are used to convert into oil. It helps to remove the plastic and prevents the entanglement of wildlife. It is used in waterways to prevent plastic from reaching the ocean [71, 72].

According to the Clean Ocean's project, plastic waste is collected from the ocean gyres using collection vessels and converted into liquid hydrocarbon fuel using the thermal degradation method. This method is applicable for the low and high-density plastics, which are heated up to 430 °C and converted into liquid hydrocarbon fuel without causing smoke and produce only less waste [73 - 75].

Advanced technologies such as radiofrequency identification (RFID) tags and cellular transmitters could be used to track and reduce the impacts of plastics in the ecosystem [76]. Plastic recycling and reuse minimize the effect of waste in landfills and the environment.

Manta trawlers are used for the collection of pollutants from the ocean for data analysis [77]. It should be kept behind the boat that skims the surface of the water and collects the floating plastic debris. The ocean currents are studied by satellite-tracked Lagrangian drifters that locate the floating marine debris [78, 79]. The study of tracking plastic debris is useful to identify the environment that is most vulnerable. It is useful to protect marine biota from the threatening effects of plastics by taking immediate action.

The effects of plastics on the environment are minimized by the following measures:

1. Reduce the use of plastics in our everyday life.
2. Recycle and reuse the plastics in an ecofriendly way.
3. Using and throwing items made of plastics should be avoided.
4. Use jute/cloth bags for shopping to avoid plastic waste.
5. Use only biodegradable plastics, which are eco-safe and also safe to organisms.
6. Avoid the use of plastics as packaging material.
7. The policies related to the control of plastic and marine pollution should be implemented and followed properly.
8. The water resource is very precious and should not be contaminated with plastic debris. Hence the plastic waste should be collected at the source itself and use only ecofriendly methods for efficient management.

**CONSENT FOR PUBLICATION**

Not applicable.

**CONFLICT OF INTEREST**

The authors confirm that this chapter contents have no conflict of interest.

## ACKNOWLEDGEMENTS

Declared none.

## REFERENCES

[1]     Derraik JGB. The pollution of the marine environment by plastic debris: a review. Mar Pollut Bull 2002; 44(9): 842-52.
[http://dx.doi.org/10.1016/S0025-326X(02)00220-5] [PMID: 12405208]

[2]     Lattin GL, Moore CJ, Zellers AF, Moore SL, Weisberg SB. A comparison of neustonic plastic and zooplankton at different depths near the southern California shore. Mar Pollut Bull 2004; 49(4): 291-4.
[http://dx.doi.org/10.1016/j.marpolbul.2004.01.020] [PMID: 15341821]

[3]     Allsopp M, Walters A, Santillo D, Johnston P. Plastic Debris in the World Oceans. Greenpeace Report, Netherlands 2006.

[4]     Arthur C, Baker J, Bamford H. Proceedings of the international research workshop on the occurrence, Effects and Fate of Microplastic Marine Debris. NOAA Technical Memorandum NOS-QR& R-30 Sept 9-11 2008.

[5]     Browne MA, Galloway T, Thompson R. Microplastic--an emerging contaminant of potential concern? Integr Environ Assess Manag 2007; 3(4): 559-61.
[http://dx.doi.org/10.1002/ieam.5630030412] [PMID: 18046805]

[6]     Gregory MR, Andrady AL. Plastic in the marine environment.Plastics and the environment. New York: John Wiley 2003; pp. 379-401.

[7]     Moore CJ. Synthetic polymers in the marine environment: a rapidly increasing, long-term threat. Environ Res 2008; 108(2): 131-9.
[http://dx.doi.org/10.1016/j.envres.2008.07.025] [PMID: 18949831]

[8]     Leslie H, van der Meulen MD, Kleissen FM, Vethaak AD. Microplastic Litter in the Dutch Marine Environment – Providing facts and analysis for Dutch policymakers concerned with marine microplastic litter. the Netherlands: Deltares 2011.

[9]     Gordon M. Eliminating Land-based Discharges of Marine Debris.California: A Plan of Action from The Plastic Debris Project. Sacramento, CA: California State Water Resources Control Board 2006.

[10]    Teuten EL, Rowland SJ, Galloway TS, Thompson RC. Potential for plastics to transport hydrophobic contaminants. Environ Sci Technol 2007; 41(22): 7759-64.
[http://dx.doi.org/10.1021/es071737s] [PMID: 18075085]

[11]    Barnes DKA. Drifting plastic and its consequences for sessile organism dispersal in the Atlantic Ocean. Mar Biol 2005; 146: 815-25.
[http://dx.doi.org/10.1007/s00227-004-1474-8]

[12]    Koelmans AA, Bakir A, Burton GA, Janssen CR. Microplastic as a vector for chemicals in the aquatic environment: critical review and model-supported reinterpretation of empirical studies. Environ Sci Technol 2016; 50(7): 3315-26.
[http://dx.doi.org/10.1021/acs.est.5b06069] [PMID: 26946978]

[13]    Teuten EL, Saquing JM, Knappe DR, *et al.* Transport and release of chemicals from plastics to the environment and to wildlife. Philos Trans R Soc Lond B Biol Sci 2009; 364(1526): 2027-45.
[http://dx.doi.org/10.1098/rstb.2008.0284] [PMID: 19528054]

[14]    Thompson RC, Moore CJ, vom Saal FS, Swan SH. Plastics, the environment and human health: current consensus and future trends. Philos Trans R Soc Lond B Biol Sci 2009; 364(1526): 2153-66.
[http://dx.doi.org/10.1098/rstb.2009.0053] [PMID: 19528062]

[15]    Hahladakis JN, Velis CA, Weber R, Iacovidou E, Purnell P. An overview of chemical additives present in plastics: Migration, release, fate and environmental impact during their use, disposal and

recycling. J Hazard Mater 2018; 344: 179-99.
[http://dx.doi.org/10.1016/j.jhazmat.2017.10.014] [PMID: 29035713]

[16]    von Moos N, Burkhardt-Holm P, Köhler A. Uptake and effects of microplastics on cells and tissue of the blue mussel Mytilus edulis L. after an experimental exposure. Environ Sci Technol 2012; 46(20): 11327-35.
[http://dx.doi.org/10.1021/es302332w] [PMID: 22963286]

[17]    Corcoran PL, Biesinger MC, Grifi M. Plastics and beaches: a degrading relationship. Mar Pollut Bull 2009; 58(1): 80-4.
[http://dx.doi.org/10.1016/j.marpolbul.2008.08.022] [PMID: 18834997]

[18]    Shaw DG, Day RH. Colour- and form- dependent loss of plastic microdebris from the North Pacific Ocean. Mar Pollut Bull 1994; 28(1): 39-43.
[http://dx.doi.org/10.1016/0025-326X(94)90184-8]

[19]    Greene CH. Planktivore functional groups and patterns of prey selection in pelagic communities. J Plankton Res 1985; 7(1): 35-40.
[http://dx.doi.org/10.1093/plankt/7.1.35]

[20]    Fossi MC, Panti C, Guerranti C, *et al.* Are baleen whales exposed to the threat of microplastics? A case study of the Mediterranean fin whale (*Balaenoptera physalus*). Mar Pollut Bull 2012; 64(11): 2374-9.
[http://dx.doi.org/10.1016/j.marpolbul.2012.08.013] [PMID: 22964427]

[21]    Muthukumar T, Aravinthan A, Lakshmi K, Venkatesan R, Vedaprakash Land Doble M. Fouling and stability of polymers and composites in marine environment. Int Biodeterior Biodegradation 2011; 65(2): 276-84.
[http://dx.doi.org/10.1016/j.ibiod.2010.11.012]

[22]    Andrady AL. Microplastics in the marine environment. Mar Pollut Bull 2011; 62(8): 1596-605.
[http://dx.doi.org/10.1016/j.marpolbul.2011.05.030] [PMID: 21742351]

[23]    McMahon CR, Hooley D, Robinson S. The diet of itinerant male Hooker's sea lions, Phocarctos hookeri, at sub-Antarctic Macquarie Island. Wildl Res 1999; 26(6): 839-46.
[http://dx.doi.org/10.1071/WR98079]

[24]    Goldsworthy SD, Hindell MA, Crowley HM. Diet and diving behaviour of sympatric fur seals Arctocephalus gazella and A. tropicalis at Macquarie Island.Marine Mammal Research in the Southern Hemisphere Status, Ecology and Medicine. New South Wales, Australia: Surrey Beatty & Sons 1997; Vol. 1: pp. 151-63.

[25]    Ogata Y, Takada H, Mizukawa K, *et al.* International Pellet Watch: global monitoring of persistent organic pollutants (POPs) in coastal waters. 1. Initial phase data on PCBs, DDTs, and HCHs. Mar Pollut Bull 2009; 58(10): 1437-46.
[http://dx.doi.org/10.1016/j.marpolbul.2009.06.014] [PMID: 19635625]

[26]    Auman HJ, Woehler EJ, Riddle MJ, Burton H. First evidence of ingestion of plastic debris by seabirds at sub-antarctic heard island. Mar Ornithol 2004; 32: 105-6.

[27]    Boerger CM, Lattin GL, Moore SL, Moore CJ. Plastic ingestion by planktivorous fishes in the North Pacific Central Gyre. Mar Pollut Bull 2010; 60(12): 2275-8.
[http://dx.doi.org/10.1016/j.marpolbul.2010.08.007] [PMID: 21067782]

[28]    Lusher AL, McHugh M, Thompson RC. Occurrence of microplastics in the gastrointestinal tract of pelagic and demersal fish from the English Channel. Mar Pollut Bull 2013; 67(1-2): 94-9.
[http://dx.doi.org/10.1016/j.marpolbul.2012.11.028] [PMID: 23273934]

[29]    Cole M, Lindeque P, Fileman E, *et al.* Microplastic ingestion by zooplankton. Environ Sci Technol 2013; 47(12): 6646-55.
[http://dx.doi.org/10.1021/es400663f] [PMID: 23692270]

[30]    Laist DW. Overview of the biological effects of lost and discarded plastic debris in the marine

environment. Mar Pollut Bull 1987; 18: 319-26.
[http://dx.doi.org/10.1016/S0025-326X(87)80019-X]

[31]  Browne MA, Dissanayake A, Galloway TS, Lowe DM, Thompson RC. Ingested microscopic plastic translocates to the circulatory system of the mussel, *Mytilus edulis* (L). Environ Sci Technol 2008; 42(13): 5026-31.
[http://dx.doi.org/10.1021/es800249a] [PMID: 18678044]

[32]  Azzarello M. Yand Vleet ES. Marine birds and plastic pollution. Mar Ecol Prog Ser 1987; 37: 295-303.
[http://dx.doi.org/10.3354/meps037295]

[33]  Setälä O, Fleming-Lehtinen V, Lehtiniemi M. Ingestion and transfer of microplastics in the planktonic food web. Environ Pollut 2014; 185: 77-83.
[http://dx.doi.org/10.1016/j.envpol.2013.10.013] [PMID: 24220023]

[34]  Rummel CD, Löder MGJ, Fricke NF, *et al.* Plastic ingestion by pelagic and demersal fish from the North Sea and Baltic Sea. Mar Pollut Bull 2016; 102(1): 134-41.
[http://dx.doi.org/10.1016/j.marpolbul.2015.11.043] [PMID: 26621577]

[35]  Browne MA, Galloway T, Thompson R. Microplastic--an emerging contaminant of potential concern? Integr Environ Assess Manag 2007; 3(4): 559-61.
[http://dx.doi.org/10.1002/ieam.5630030412] [PMID: 18046805]

[36]  Jacobsen JK, Massey L, Gulland F. Fatal ingestion of floating net debris by two sperm whales (*Physeter macrocephalus*). Mar Pollut Bull 2010; 60(5): 765-7.
[http://dx.doi.org/10.1016/j.marpolbul.2010.03.008] [PMID: 20381092]

[37]  van Franeker JA, Blaize C, Danielsen J, *et al.* Monitoring plastic ingestion by the northern fulmar *Fulmarus glacialis* in the North Sea. Environ Pollut 2011; 159(10): 2609-15.
[http://dx.doi.org/10.1016/j.envpol.2011.06.008] [PMID: 21737191]

[38]  Villiers MS, Bruyn PJN. Stone-swallowing by three species of Penguins atsub-antarctic Marion Island. Mar Ornithol 2004; 32: 185-6.

[39]  Mascarenhas R, Santos R, Zeppelini D. Plastic debris ingestion by sea turtle in Paraíba, Brazil. Mar Pollut Bull 2004; 49(4): 354-5.
[http://dx.doi.org/10.1016/j.marpolbul.2004.05.006] [PMID: 15341830]

[40]  Mrosovsky N, Ryan GD, James MC. Leatherback turtles: the menace of plastic. Mar Pollut Bull 2009; 58(2): 287-9.
[http://dx.doi.org/10.1016/j.marpolbul.2008.10.018] [PMID: 19135688]

[41]  Luís LG, Ferreira P, Fonte E, Oliveira M, Guilhermino L. Does the presence of microplastics influence the acute toxicity of chromium(VI) to early juveniles of the common goby (*Pomatoschistus microps*)? A study with juveniles from two wild estuarine populations. Aquat Toxicol 2015; 164: 163-74.
[http://dx.doi.org/10.1016/j.aquatox.2015.04.018] [PMID: 26004740]

[42]  Barboza LGA, Gimenez BCG. Microplastics in the marine environment: Current trends and future perspectives. Mar Pollut Bull 2015; 97(1-2): 5-12.
[http://dx.doi.org/10.1016/j.marpolbul.2015.06.008] [PMID: 26072046]

[43]  Avio CG, Gorbi S, Milan M, *et al.* Pollutants bioavailability and toxicological risk from microplastics to marine mussels. Environ Pollut 2015; 198: 211-22.
[http://dx.doi.org/10.1016/j.envpol.2014.12.021] [PMID: 25637744]

[44]  Cheung LTO, Lui CY, Fok L. Microplastic contamination of wild and captive flathead Grey mullet (*Mugil cephalus*). Int J Environ Res Public Health 2018; 15(4): 597.
[http://dx.doi.org/10.3390/ijerph15040597] [PMID: 29587444]

[45]  Bråte ILN, Eidsvoll DP, Steindal CC, Thomas KV. Plastic ingestion by Atlantic cod (*Gadus morhua*) from the Norwegian coast. Mar Pollut Bull 2016; 112(1-2): 105-10.
[http://dx.doi.org/10.1016/j.marpolbul.2016.08.034] [PMID: 27539631]

[46]    Bessa F, Barría P, Neto JM, *et al.* Occurrence of microplastics in commercial fish from a natural estuarine environment. Mar Pollut Bull 2018; 128: 575-84.
[http://dx.doi.org/10.1016/j.marpolbul.2018.01.044] [PMID: 29571409]

[47]    Renzi M, Guerranti C, Blašković A. Microplastic contents from maricultured and natural mussels. Mar Pollut Bull 2018; 131(Pt A): 248-51.
[http://dx.doi.org/10.1016/j.marpolbul.2018.04.035] [PMID: 29886944]

[48]    Lusher AL, Hollman PCH, Mendoza-Hill JJ. Microplastics in fisheries and aquaculture: Status of knowledge on their occurrence and implications for aquatic organisms and food safety. Rome, Italy: Food & Agriculture Organization 2017.

[49]    Sources, fate and effects of microplastics in the marine environment: Part two of a global assessment. IMO/FAO/ UNESCO- IOC/UNIDO/WMO/IAEA/UN/UNEP/UNDP Joint Group of Experts on the Scientific Aspects of Marine Environmental Protection) Rep Stud GESAMP No 93. 2016.

[50]    Rochman CM, Tahir A, Williams SL, *et al.* Anthropogenic debris in seafood: Plastic debris and fibers from textiles in fish and bivalves sold for human consumption. Sci Rep 2015; 5: 14340.
[http://dx.doi.org/10.1038/srep14340] [PMID: 26399762]

[51]    Karami A, Golieskardi A, Ho YB, Larat V, Salamatinia B. Microplastics in eviscerated flesh and excised organs of dried fish. Sci Rep 2017; 7(1): 5473.
[http://dx.doi.org/10.1038/s41598-017-05828-6] [PMID: 28710445]

[52]    Bellas J, Martínez-Armental J, Martínez-Cámara A, Besada V, Martínez-Gómez C. Ingestion of microplastics by demersal fish from the Spanish Atlantic and Mediterranean coasts. Mar Pollut Bull 2016; 109(1): 55-60.
[http://dx.doi.org/10.1016/j.marpolbul.2016.06.026] [PMID: 27289284]

[53]    Abbasi S, Soltani N, Keshavarzi B, Moore F, Turner A, Hassanaghaei M. Microplastics in different tissues of fish and prawn from the Musa Estuary, Persian Gulf. Chemosphere 2018; 205: 80-7.
[http://dx.doi.org/10.1016/j.chemosphere.2018.04.076] [PMID: 29684694]

[54]    Andrady AL. The plastic in microplastics: A review. Mar Pollut Bull 2017; 119(1): 12-22.
[http://dx.doi.org/10.1016/j.marpolbul.2017.01.082] [PMID: 28449819]

[55]    RASFF Rapid Alert System for Food and Feed (RASFF). 2015. https://webgate.ec.europa.eu/rasff-window/portal/?event=searchResultList

[56]    EFSA CONTAM Panel (EFSA Panel on Contaminants in the Food Chain).. Statement on the presence of microplastics and nanoplastics in food, with particular focus on seafood.EFSA J 2016(14): 4501. 30p

[57]    Liebezeit G, Liebezeit E. Non-pollen particulates in honey and sugar. Food Addit Contam Part A Chem Anal Control Expo Risk Assess 2013; 30(12): 2136-40.
[http://dx.doi.org/10.1080/19440049.2013.843025] [PMID: 24160778]

[58]    Liebezeit G, Liebezeit E. Synthetic particles as contaminants in German beers. Food Addit Contam Part A Chem Anal Control Expo Risk Assess 2014; 31(9): 1574-8.
[http://dx.doi.org/10.1080/19440049.2014.945099] [PMID: 25056358]

[59]    Kosuth M, Mason SA, Wattenberg EV. Anthropogenic contamination of tap water, beer, and sea salt. PLoS One 2018; 13(4): e0194970.
[http://dx.doi.org/10.1371/journal.pone.0194970] [PMID: 29641556]

[60]    Yang D, Shi H, Li L, Li J, Jabeen K, Kolandhasamy P. Microplastic pollution in table salts from China. Environ Sci Technol 2015; 49(22): 13622-7.
[http://dx.doi.org/10.1021/acs.est.5b03163] [PMID: 26486565]

[61]    Karami A, Golieskardi A, Keong Choo C, Larat V, Galloway TS, Salamatinia B. The presence of microplastics in commercial salts from different countries. Sci Rep 2017; 7: 46173.
[http://dx.doi.org/10.1038/srep46173] [PMID: 28383020]

[62]    Ashton K, Holmes L, Turner A. Association of metals with plastic production pellets in the marine environment. Mar Pollut Bull 2010; 60(11): 2050-5.
[http://dx.doi.org/10.1016/j.marpolbul.2010.07.014] [PMID: 20696443]

[63]    Bakir A, Rowland SJ, Thompson RC. Competitive sorption of persistent organic pollutants onto microplastics in the marine environment. Mar Pollut Bull 2012; 64(12): 2782-9.
[http://dx.doi.org/10.1016/j.marpolbul.2012.09.010] [PMID: 23044032]

[64]    Koelmans AA, Besseling E, Foekema EM. Leaching of plastic additives to marine organisms. Environ Pollut 2014; 187: 49-54.
[http://dx.doi.org/10.1016/j.envpol.2013.12.013] [PMID: 24440692]

[65]    Holmes LA, Turner A, Thompson RC. Adsorption of trace metals to plastic resin pellets in the marine environment. Environ Pollut 2012; 160(1): 42-8.
[http://dx.doi.org/10.1016/j.envpol.2011.08.052] [PMID: 22035924]

[66]    Eagles-Smith CA, Silbergeld EK, Basu N, *et al.* Modulators of mercury risk to wildlife and humans in the context of rapid global change. Ambio 2018; 47(2): 170-97.
[http://dx.doi.org/10.1007/s13280-017-1011-x] [PMID: 29388128]

[67]    Zettler ER, Mincer TJ, Amaral-Zettler LA. Life in the "plastisphere": microbial communities on plastic marine debris. Environ Sci Technol 2013; 47(13): 7137-46.
[http://dx.doi.org/10.1021/es401288x] [PMID: 23745679]

[68]    Keswani A, Oliver DM, Gutierrez T, Quilliam RS. Microbial hitchhikers on marine plastic debris: Human exposure risks at bathing waters and beach environments. Mar Environ Res 2016; 118: 10-9.
[http://dx.doi.org/10.1016/j.marenvres.2016.04.006] [PMID: 27128352]

[69]    Viršek MK, Lovšin MN, Koren Š, Kržan A, Peterlin M. Microplastics as a vector for the transport of the bacterial fish pathogen species *Aeromonas salmonicida.* Mar Pollut Bull 2017; 125(1-2): 301-9.
[http://dx.doi.org/10.1016/j.marpolbul.2017.08.024] [PMID: 28889914]

[70]    Schirinzi GF, Pérez-Pomeda I, Sanchís J, Rossini C, Farré M, Barceló D. Cytotoxic effects of commonly used nanomaterials and microplastics on cerebral and epithelial human cells. Environ Res 2017; 159: 579-87.
[http://dx.doi.org/10.1016/j.envres.2017.08.043] [PMID: 28898803]

[71]    The Ocean Clean-Up. 2014.http://www.theoceancleanup.com/the-concept.html

[72]    Singh T. 19-year-old develops ocean cleanup array that could remove 7,250,000 tons of plastic from the world's oceans 2013. http://inhabitat.com/19- year-old-student-develops-ocean-cleanup-arr-y-that-could remove- 7250000-tons-of-plastic-from-the-worlds-oceans/

[73]    Livingeco The Clean Oceans Project/Plastic to Oil Machine, 2011. http://www. Youtube.com/watch?v=8qBFlOqLnJ8

[74]    Sarker M, Rashid MM, Molla M, Sadikur Rahman M. A new technology proposed to recycle waste plastics into hydrocarbon fuel in USA. International Energy Environ 2012; 3(5): 749-60.

[75]    Sarker M, Rashid MM, Molla M, Rahman MS. Thermal conversion of waste plastics (HDPE, PP and PS) to produce mixture of hydrocarbons. American J Environ Engineering 2012; 2(5): 128-36.
[http://dx.doi.org/10.5923/j.ajee.20120205.03]

[76]    Greengard S. Tracking garbage. Commun ACM 2010; 53(3): 19-20.
[http://dx.doi.org/10.1145/1666420.1666429]

[77]    Ryan PG, Moore CJ, van Franeker JA, Moloney CL. Monitoring the abundance of plastic debris in the marine environment. Philosophical Transactions of the Royal Society, B. Biol Sci 2009; 364(1526): 1999-2012.
[http://dx.doi.org/10.1098/rstb.2008.0207]

[78]    Maximenko N, Hafner J, Niiler P. Pathways of marine debris derived from trajectories of Lagrangian drifters. Mar Pollut Bull 2012; 65(1-3): 51-62.

[http://dx.doi.org/10.1016/j.marpolbul.2011.04.016] [PMID: 21696778]

[79]   Martinez E, Maamaatuaiahutapu K, Taillandier V. Floating marine debris surface drift: convergence and accumulation toward the South Pacific subtropical gyre. Mar Pollut Bull 2009; 58(9): 1347-55. [http://dx.doi.org/10.1016/j.marpolbul.2009.04.022] [PMID: 19464033]

# CHAPTER 3

# Thermal Pollution

**M. Sudha Devi**[1,*] and **T. Dhanalakshimi**[2]

*¹ Department of Biochemistry, Biotechnology and Bioinformatics, School of Biosciences, Avinashilingam Institute for Home Science and Higher Education for Women, Coimbatore 641 043, Tamil Nadu, India*

*² Department of Biology, Ministry of Education, Republic of Maldives*

**Abstract:** Pollution is the contamination of the environment by synthetic, semi-synthetic or unreal substances or energy that adversely impacts living or non-living matter interfering with health, quality of life or the normal functioning of the ecosystems. To put it in a nutshell, it is the presence of the wrong substance in the wrong place in the wrong quantities at the wrong time. The WordWeb dictionary defines 'pollutants' as waste matters that contaminate water, air, or soil. Temperature is a property of a system which determines if they are in thermal equilibrium; temperature plays a vital role in determining the condition in which living things can survive. Thermal pollution can be defined as 'the warming of the ecosystem due to which the desirable living conditions of organisms are adversely impacted. Studies reveal even a minor change in ambient temperature and oxygen levels can have a profound effect on ecosystems. When an industry/organization consumes water from a natural source, it discharges the water back into natural resources upon heating it up or cooling it down, which subsequently changes the oxygen levels; eventually, it has devastative effects on local ecosystems and populations. It is high time to establish proper conservation and management measures to tide over the crisis of pollution. Upon extensive studies, it has been found that the Three 'R's - Reduce, Replace, Reuse, can be of great help to tackle pollution. Let us explore and elaborately discuss thermal pollution. Firstly, the classification of pollution, causes, effects and probable solutions for thermal pollution are also discussed .

**Keywords:** Classification, Effects, Pollution Management, Thermal Pollution, Warming Ecosystem.

## INTRODUCTION

### Definition

Thermal pollution can be defined as an abrupt change in the temperature of a

---

* **Corresponding author M. Sudha Devi:** Department of Biochemistry, Biotechnology and Bioinformatics, Avinashilingam Institute for Home Science and Higher Education for Women, Coimbatore - 641 043, Tamil Nadu, India; Tel: +91 90039 54969; E-mail: sudhadevi2003@gmail.com

**J. Senthil Kumar, P. Ponmurugan & A. Vinothkanna (Eds.)**

natural water body like an ocean, lake, river or pond due to human influence. It is not only caused by hot water, but also by the cold water that is discharged by various industries into the rivers or seas containing warm water. The significant concern that is raised nowadays is the implacable adversity called thermal pollution. In the name of development, human beings are interfering with the cycle of nature and have made our planet 'earth' suffer from various hazardous problems. Thermal pollution is one such problem that desperately needs a solution. It is the harmful discharge of heated/cooled liquid into water-bodies or heat released into the atmosphere as a waste product of any industry or nuclear power plant. It normally occurs when a plant or facility takes in water from a natural resource and discharges it back with an altered temperature that causes a rise in the water temperature, which subsequently affects the ecological balance. Usually, these facilities consume water for cooling their machinery or to help better produce their products, *e.g.*, water is used for cooling in a power plant that runs into a nearby river, and when it is discharged back into the river, it drastically affects the river's ecosystem.

Thermal pollution is one parameter of the broader subject of water pollution. The cause of it is the use of water as a coolant by industrial manufacturers and power plants. It is a result of discharging heated water back into the water bodies, the addition of surplus heat to water and ejecting it back. This change in temperature decreases oxygen supply and affects ecosystem composition. This condition primarily arises from the waste heat generated by an industrial process like power generation plants.

Studies say thermal or molecular power generation plants are primarily responsible for contributing about 70-80% discharge of heated water to aquatic bodies. Heated waters from these sources have 10-14°C higher temperatures as compared to the temperatures of the receiving waters. Besides, the sewage effluents and wastewaters have a 4-8°C higher temperature than the aquatic bodies where they are discharged.

## Causes

Water works as a cooling agent in power, manufacturing and industrial plants. Production and manufacturing plants cause major thermal pollution. The water discharged by these plants into the ocean raises the temperature sharply. When oxygen levels are altered in the water, this can also degrade the quality and longevity of life in wildlife that lives underwater. This process can also wipe away riverside vegetation, which constantly depends on constant levels of oxygen and temperature. By altering these natural environments, industries are essentially helping decrease the quality of life for these marine based life forms and can ultimately destroy habitats if they are not controlled and careful about their practices. These plants use water to cool down their machines and eject back the hot water into water bodies. Thus, the natural water goes through a sudden rise in temperature. Fig. (**1**) gives a clear picture of global thermal pollution of rivers from thermoelectric power plants [1].

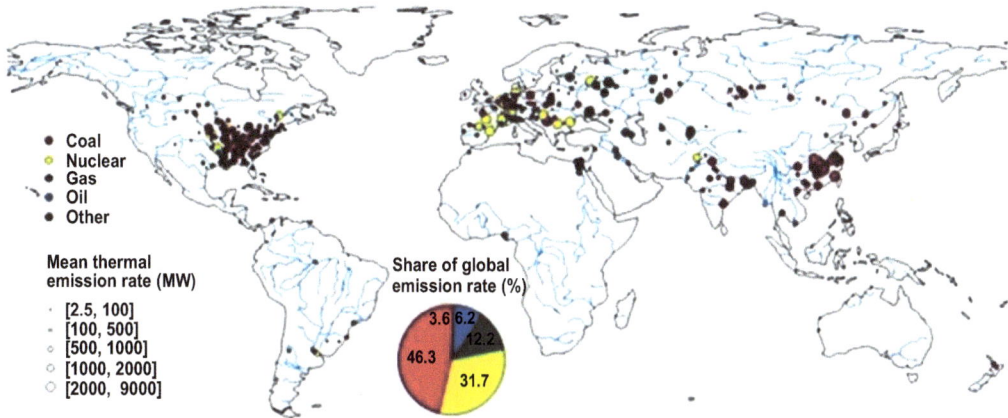

**Fig.(1).** Global thermal pollution of rivers from thermoelectric power plants [1].

Along the riversides or bank of the rivers, many ancient civilizations flourished to name Indus Valley the river Saraswati, Vedic or Harappan Civilization; Pativilca, Fortaleza, and Supe the Norte Chico Civilization; the Nile the Ancient Egypt Civilization [2]. But the modern, industrial development is continued to exploit and pollute the water bodies.

## *Industrial Effluents*

Drainage from research institutions, hospitals, sugar industries, and nuclear experiments, power planets and explosives emits a lot of unutilized heat and traces of heat-absorbing materials into neighboring water resources resulting in raised temperatures of the water bodies with a deep decrease in the levels of dissolved oxygen. Eventually, it causes ecocide in aquatic life [3].

## *Domestic Sewages*

A common phenomenon in almost all parts of the world is the discharge of domestic sewage into lakes, rivers, streams or canals, often with minimal treatment or without treatment at all, unaware of the fact that the sewage water has a higher temperature than the receiving water. Wastewater has a high organic load that alters the temperature of receiving water leading to a reduction of dissolved oxygen content and increased anaerobic conditions, eventually, resulting in death of various forms of aquatic life [4].

## Soil Erosion

One of the major causes of thermal pollution is consistent soil erosion in water bodies. It makes them more vulnerable to sunlight. The high temperature could prove fatal for aquatic biomass as it may generate anaerobic conditions. Soil erosion can result in thermal pollution, causing siltation and sedimentation, raising the water levels that enlarge the surface area of the water exposed to sunlight. Thus, increased water temperatures cause thermal pollution. Streamside erosion may also remove vegetation cover along the streams enlarging the area of exposure to direct sunlight. This temperature increase can have destructive effects on the aquatic community [5].

## Deforestation

When the trees and plants are removed, this results in the direct exposure of sunlight over water bodies and they absorb more heat and raise its temperature. It is a major cause of the higher concentrations of greenhouse gases, *i.e.* global warming. Forests and vegetation cover is responsible for absorbing the sun's heat and reflecting it. It also combats the accumulation of intense heat on rivers, ponds, canals, and lakes. It is clear that deforestation leaves water sources directly exposed to sunlight and allows extreme heat of sunlight that would have otherwise been absorbed by the vegetation cover to accumulate in water sources. In addition to this, atmospheric temperatures are heightened due to the thermal effect of $CO_2$ greenhouse gases, which could have been absorbed by trees. Heightened atmospheric temperatures mean compounded temperature levels of the water resources [6].

## Urban Stormwater Runoff

During hot summer days and months, urban surfaces, including parking lots, macadam roads, and sidewalks, usually heat up by absorbing the thermal radiation from the sun. The urban stormwater runoff absorbs the heat while raining and discharges it into the nearby sewer systems and water bodies. The mean temperature of the water is altered due to the urban runoff discharged to surface waters from paved surfaces like roads and parking lots. During summer seasons, the pavement gets quite hot, which creates a warm runoff that gets into the sewer systems and water bodies. Many urban areas like parking places, roads, *etc.*, deposit rain water and discharge the heated water back into water bodies. The heated water alters the normal temperature of natural water bodies [7].

## *Natural Causes*

Geothermal activity and volcanoes under the seas and oceans can trigger warm lava to raise the temperature of water bodies and also can induce natural underground heating of the water bodies. Similarly, lightning can also add a massive amount of heat into the water bodies, resulting in an increase in the overall temperature of the water source, having profound impacts on the environment. Hot rocks and volcanic lava have the potential of heating and raising water body temperatures. The rise in temperatures merely translates to significant detrimental impacts on marine life and the environment, in general. Thus, natural geothermal activities can stimulate lava and can cause a rise in water temperature, paving way for thermal pollution [8].

## *Hydroelectric Power Generation*

It leads to negative thermal loading in water resources. The turning of steam-heated turbines, heats up the water and dumps it back into the receiving water at higher temperatures. Consequently, it affects the seasonal activities, diurnal and also the metabolic responses of organisms in the water resources used to generate power [9].

## *Livestock Waste Mixed into Water*

This is yet another cause of thermal pollution. Many industries dispose their untreated livestock wastes into water without analyzing the hazardous consequences of this act [9].

## *Release of Cold Water*

Many industries discharge cool water from their reservoirs to warm water rivers, lakes or ponds that creates an imbalance in flora and fauna of affected water bodies [10].

## *Unawareness Among People*

Growing thermal pollution is also due to a lack of awareness among people. Despite knowing the hazardous effects of thermal pollution on environment, there are abundant industries which are continuously using ways that cause thermal pollution [4].

## EFFECTS

There are generally two schools of thought to combat the harmful effects of thermal pollution i) Negative effects of this pollution on marine ecosystems and how detrimental it is to positive environmental practices ii) some of the most basic parts of human life would be completely obsolete without these industries operating the way they do. Wastewater cannot be properly maintained, we would have no industries that could produce the goods we need, and so on. The effects of thermal pollution on ecosystems, however, greatly outweigh the benefits that industries have by participating in the act. With an increase in temperature of the aquatic bodies, the rate of exchange of salts in the organism increases, and any toxin is liable to exert greater effects and temperature fluctuations are likely to affect organisms. Other adverse effects of aquatic pollution on aquatic life include (i) early hatching of fish eggs (ii) increase in biological oxygen demand (iii) change in diurnal and seasonal behaviour and metabolic response of organisms and (iv) decrease in species diversity [9].

The harmful effects of thermal pollution are discussed below:

### Reduction in Dissolved Oxygen (DO)

The warm temperature reduces the levels of DO in water. Compared to cold water, warm water holds pretty less oxygen creating suffocation for plants and animals such as fish, copepods and amphibians, which results in anaerobic conditions. Besides this, warmer water permit algae to propagate on the surface of water and over the long term, growing algae can decrease oxygen levels in the water. The pollutants from various industrial plants are heated, this decreases the availability of oxygen with an increase in the temperature of water [3, 9].

### Reduced Solubility of Oxygen

This is another fatal effect of thermal pollution. This less solubility of oxygen in water significantly affects the metabolism of water animals [3, 4].

### Change in Water Properties

The decline in viscosity, density and solubility of gases in water increases the setting speed of suspended particles which seriously affects the food supplies of aquatic organism [3, 4].

### Increase in Toxicity

The increased level of pollutants causes an increase in the temperature of water, which amplifies the toxicity of the poison present in water. There are plentiful factories that discharge their chemical waste directly into water bodies. This not only causes thermal pollution but also makes the water poisonous. The toxicity in water increases the death rate in marine life. With the regular flow of water with elevated temperature discharge from industries, there is a huge raise in toxins that are being regurgitated into the natural body of water. These toxins may contain radiation or chemicals that may have a cruel impact on the local ecology and make them susceptible to various diseases [4].

## Disruption of Biological Activities

Temperature changes disturb the whole marine ecosystem since, temperature has a strong impact on metabolism, physiology and biological processes like respiration rate, digestion, excretion and development of an aquatic organism [3].

## Damage of Biotic Organism

Thermal pollution may cause a significant loss of biodiversity. Organisms that can adapt to the temperature change may have an advantage over organisms that are not used to the warmer temperatures. Changes in the environment may cause certain species of organisms to shift theirbase to an alternate place, while there could be a significant number of species that may shift in because of warmer waters. Aquatic organisms like juvenile fish, plankton, fish, eggs, larva, algae and protozoa, which pass through screens and condenser cooling systems are extremely sensitive to abrupt temperature changes. When the marine lives undergo a sudden increase or decrease in temperature of water bodies, it destroys the life of aquatic organisms. Lethal temperature for trout is 22°C, for yellow perch 35°C and for carp, it is 32°C. Heat affects the fish nervous system, inactivates the enzyme and coagulates cell protoplasm that results in the thermal death of fish [5].

## Ecological Impact

A thermal shock can trigger the mass death of fish, insects, amphibians or plants. It could be lethal for some species, while hotter water may prove favorable for other species. Minimal water temperature increases the level of activity, while elevated temperature decreases the level of activity. Many aquatic species are sensitive to even one degree Celsius raise that can cause a significant impact on their metabolism and other cellular activities [10].

## Affects Reproductive Systems

Increasing temperatures can result in arresting the reproduction of marine wildlife as reproduction can happen within a particular range of temperatures. Extreme temperature can cause the release of immature eggs or can prevent the normal development of eggs [9].

## Increases Metabolic Rate

An increase in the metabolic rate is observed in organisms due to the thermal pollution by escalating enzyme activity, which causes organisms to consume more food more than required when their environment is not changed. This, in turn, interrupts the food chain and the balance of the species composition is altered. It has been determined that a temperature rise of 10°C will double the rate of many chemical reactions and cause the decomposition of the organic matter. On the other hand, an increase in temperature leads to an increase in the metabolism of organisms. Cold-blooded animals, especially fishes, are exceptionally sensitive to temperature changes and they will fade away if heated effluents are discharged in it. The living organisms within the aquatic bodies are adversely affected by the added heat. With the increase of temperature, the oxygen consumption and heart rate of fish increase to obtain oxygen for increased metabolic processes. On the other hand, the oxygen concentration of water decreases and hence fishes struggle for their survival. Some plants and animals die due to exposure to hot water [3, 9].

## Migration

Change in the temperature of water can also cause some selective species of organisms to migrate to a suitable environment that favors their survival requirements, resulting in the loss of those species that solely depend on them for their daily food as their food chain is interrupted. Species intolerant to higher temperatures migrate to lower layers, which are cooler and there are many planktonic forms which settle down at the bottom layer. Perhaps the heated water may contain some organic matter such as organic acids, hydrogen sulphide and ammonia, so it damages plant and animal life mainly due to the depletion of oxygen [3, 4, 9].

## Adverse Effect on Water Plants

Change in temperature levels is extremely harmful to the aquatic plants. These plants cannot cope up with the sudden alteration in water temperature. Consequently, many aquatic plants are extinct each day because of thermal pollution. Especially, the most important thing to consider is the effects of thermal pollution that greatly overshadow the human need. Plants and industries have been able to find successful ways to deal with thermal pollution, but many of them are not practicing them because it is simply easier to work from the conventional model. If we want to enhance the environment that surrounds marine biology, then the attitude towards thermal pollution needs to be changed radically . By being aware of the causes and effects, one can have a significant impact on how these plants are used choose to make a change [3, 9].

## CONTROL

There is little one can do about this type of pollution except to wait and allow the waters to cool down to biologically harmless temperatures before being discharged into some water bodies. If the volume of heated water is manageable, it can be passed over a system of cascades or through fountains which cause rapid cooling. To handle large quantities of heated effluents, large tanks or reservoirs should be constructed to retain the water for a little longer time till the water is cooled down to a tolerable temperature. After which they may be released to water bodies. It is advisable to discharge the heated waters into some lotic system, wherein active churning occurs due to the flow of water, which causes rapid cooling, instead of a lentic system [3].

## PREVENTION

The following measures can be taken to curb temperature change caused by thermal pollution:

### Cooling Towers

Heated water from the industries can be properly treated by the installation of cooling ponds and cooling towers before discharging into water bodies. It is also a good idea when talking about solutions for thermal pollution. The purpose of using cooling towers is the same as artificial lakes. The cooling towers also use the hot water of industries, process it by transferring its heat and transform hot water into cold water. This cool water can be recycled and used again for different industrial purposes. Generally, the cooling towers are of two types i) dry cooling tower and ii) wet cooling tower. In dry cooling towers, the heated water is made to flow in circular, elongated pipes. Again, the cold air blows are passed upon these pipes that help in bringing down the temperature of hot water. In a wet cooling tower, the heated water gets spread upon the flow-directing panels. Afterwards, the high-speed cold air is passed upon it. Henceforth, the hot water gets cooled down [10].

### Artificial Lakes

In this man-made lake, industries can discharge their used or heated water at one end and water for cooling purposes may be withdrawn from the other end. The heat is eventually dissipated through evaporation [11].

### Saving Electricity

Power plants are the main reasons behind growing thermal pollution as they use water as a cooling agent for cooling down their machines. This used water, which is much higher in temperature, is discharged back into the rivers, seas or lakes. We can make a significant contribution to controlling thermal pollution by consuming less electricity. The use of less electricity will lead to less workload on power plants and these plants will not have to use their machines too much, suggesting controlled use of water as a coolant. Hence, switching off fans in an empty room, switch off unnecessary lights, use of solar products and techniques; all such steps will help us to consume lesser electricity [3, 12].

## Adopting Better Technologies

Science has plentiful inventions, discoveries, techniques and knowledge. Incorporating good techniques ensures a good lifestyle for the human race. The use of better technologies is strongly recommended for solving the problem of thermal pollution. There are technologies available that help in the cooling down of machines. If machines are cooled down with the help of technologies, the use of water as a coolant will come to a much reduced level. Various industries and power plants should look out for appropriate technologies that serve the purpose without increasing the ongoing problem of thermal pollution [4, 13].

## Holding Back the Water for Good

If factories or plants cannot stop using water as a coolant, there is another option available for them. After using the water as a coolant, they should store that water somewhere else for a temporary period. Instead of discharging back the heated water into water bodies, the temporarily collected heated water can be used for various other purposes too. Storing the heated water for a particular time will help in bringing back the high temperature of water to a normal level [14].

## Plantation of More Trees Upon the Banks of Rivers, Seas and other Water Bodies

This is also a good way to control thermal pollution. The trees around sources of water help in absorbing the harsh sun rays and prevent them from falling directly on the water. This helps in the prevention of heating of water bodies. Planting more trees also helps in controlling the problem of soil erosion because the strong roots of trees hold the soil firmly and prevent it from erosion. Trees not only help in controlling thermal pollution but also aid in a better environment, including fresh air and peaceful scenic views. We should also encourage our coming generations to plant more and more trees [15].

## Artificial Lakes

Industries, plants or factories which are serious about storing and reusing the heated water, used as a coolant, can work out on artificial lakes. There are artificial lakes where the heated water can be stored easily. These lakes are very helpful for normalizing the temperature of hot water. This way, the hot water will not be disposed of back to the lakes, rivers, *etc.*, and will be used in other suitable tasks. Actually, the artificial lakes or ponds use evaporation or convection techniques for cooling down the water. These artificial lakes or ponds generally contain two ends. From one end, the hot water is transferred into the lake; it is processed through evaporation or other techniques and finally, when it cools down, it is taken out from the other end. The evaporated heat dissolves in the air [8].

## Recycling Used Water

Smart people always find intelligent solutions for even the most difficult problems. If people start working upon the ideas of recycling the used water in plants and factories, the problem of thermal pollution will definitely be lessened to a significant extent. Every plant or industry should mandatorily restrict the discharge of water which was used as a coolant in the industries. Rather, it will be recycled for further tasks. In this modern era, we often hear news about the shortage of water and thousands of people are dying because of the same. Let us ask ourselves isn't it our duty to save water and use it for good [9, 16].

## Spreading Awareness Among People

The environment can be made better with united efforts. Making more and more people aware about the problem of thermal pollution will be very beneficial in the long run. Groups of people can initiate a discussion with different plants and industries. These groups can discuss the harmful effects of thermal pollution on aquatic life and our environment. We can also create awareness to people about the consistent problem of thermal pollution by gaining the right knowledge about thermal pollution [14, 17].

## Suitable Arrangements in Urban Places

Places like parking spaces, drainage pipes, sewerage tanks, *etc.*, should have proper arrangements so that the water does not get accumulated at those spaces. When the water is accumulated at these spaces, it gets heated up and gets mixed with seas, ponds, lakes, *etc.*, thus making way for thermal pollution. Hence, by making appropriate arrangements at such places, we can stop water from getting accumulated [15, 18].

## Co-generation

It is also a wonderful idea to combat thermal pollution. In the process of co-generation, the useless heat from hot water can be recycled and used smartly in many tasks by industries [19].

Hence, we can say any kind of pollution may directly or indirectly affect lives by affecting all the aspects of the environment.

## CONCLUSION

Thermal pollution is a great threat to our planet. If necessary precautions aren't taken, the problem will get increased exponentially, degrading the quality of water and adversely affecting the life on earth, ultimately affecting the humans. There is no life without water, neither for the animal nor the plants. It is so important for life. Though 70% of our planet, Earth, is covered with water, 97% of it is saltwater, which we cannot use for our drinking or other purposes. Actually, we have only 3% of freshwater, out of this, 2% of water is in the form of ice glaze, with only 1% freshwater. This fact reveals the significance of water. Hence, it is high time to take a pledge for the prevention of thermal pollution and take all necessary steps that serve this purpose. Every human can make a difference by making individual efforts. The industries, power plants and factories should also give it a serious thought. The collective effort of human beings will definitely solve the problem of thermal pollution and will help in its prevention.

## CONSENT FOR PUBLICATION

Not applicable.

## CONFLICT OF INTEREST

The authors confirm that this chapter contents have no conflict of interest.

## ACKNOWLEDGEMENTS

Declared none.

## REFERENCES

[1]     Raptis CE, Vliet MTH, Pfister S. Global thermal pollution of rivers from thermoelectric power plants 2016. https://iopscience.iop.org/article/10.1088/1748-9326/11/10/104011 [http://dx.doi.org/10.1088/1748-9326/11/10/104011]

[2]     Wiyanarti E. River and civilization in sumatera's historical perspective in the 7th to 14th centuries. IOP Conf Series: Earth and Environmental Science 2018. https://www.researchgate.net/publication/ 324947429_River_and_Civilization_in_Sumatera's_Historical_Perspective_in_The_7th_to_14th_Cent uries

[3]     Sathyanarayan S, Zade S, Sitre S, Meshram P. A Text Book of Environmental Studies. Allied Publishers Pvt. Ltd. 2009.

[4]     Asthana DK, Asthana M. Environment - Problems and Solutions. New Delhi: S. Chand and Co. Ltd. 2003.

[5]     Katyal T, Satake M. Environmental Pollution I Edition. New Delhi: Anmol Publications 1989.

[6]     Schmitz RJ. Introduction to Water Pollution Biology I Edition. Texas: Gulf Publishing Company 1996.

[7]     Purohit SS, Shammi QJ, Agrawal AK. Environmental Sciences - A Newer Approach, I Edition. India: Agrobios 2004.

[8]     Palanisamy PN, Rani MR, Manikandan P, Kowshalya VN, Geetha A. Environmental Science IV Edition. Pearson 2017.

[9]     Cunningham WP, Saigo BW. Environmental Science - A Global Concern. McGraw Hill. 2009. XI Ed.

[10]    Botkin DB, Keller EA. Environmental Science - Earth As a Living Planet VIII Edition. John Wiley & Sons Ltd. 2011.

[11]    Bailey RA, Krause S, Clark HM, Strong RL, Ferris JP. Chemistry of the Environment,. Academic Press. 1979. II Ed.

[12]    Mayer JR. Connections in Environmental Science - A case study approach. Mc Graw Hill 2000.

[13]    Agarwal KC. Environmental Biology. Bikaner: Nidi Publ. Ltd. 2001.

[14]    Brunner RC. Hazardous Waste Incineration. McGraw Hill Inc. 1989.

[15]    Cunningham WP, Cooper TH, Gorhani E, Hepworth MT. Environmental Encyclopedia, Jaico Publ House, Mumbai, India

[16]    Gleick HP. Water in crisis: A Guide to the World's Fresh Water Resources. New York, United States: Oxford University Press Inc. 1993.

[17]    Rao MN, Datta AK. Waste Water treatment. Oxford & IBH Publ. Co. Pvt. Ltd. 1987.

[18]    Sharma BK. Environmental Chemistry. Meerut: Geol Publ. House 2001.

[19]    Wanger KD. Environmental Management. Philadelphia, USA: WB Saunders Co. 1998.

# CHAPTER 4

# Agricultural Pollution

**K. Chitra**[1,*] and **B. Sathya Priya**[2]

[1] *Department of Botany, Bharathiar University, Coimbatore, Tamil Nadu, India*

[2] *Department of Environmental Sciences, Bharathiar University, Coimbatore, Tamil Nadu, India*

**Abstract:** Agriculture is a combination of science and art, and it is a complex activity. It is the cultivation of plants and livestock. One-third of the world's workers are dependent on agriculture. But in developed countries, over the centuries, the number of agricultural workers has decreased significantly. In rural areas, agriculture is the largest livelihood provider in India. The by-products of growing and raising livestock, food crops, animal feed and biofuel crops are often considered as contaminants and released into the environment, referred to as agricultural pollution. Many different sources cause agricultural pollution. They are nitrogen-based fertilizers, chemical fertilizers, pesticides, animal manure, industrial effluents, soil erosion and sedimentation, farm animal waste and nutrient runoff. Water and lakes are mainly contaminated by agricultural pollutants. Agricultural pollutants contaminate the soil and water. Chemicals from fertilizers and pesticides first contaminate the groundwater and consequently, drinking water. It contributes to health-related problems in humans. Algal blooms in drinking water and swimming in water streams containing dangerous algal blooms can cause rashes, stomach and liver problems, respiratory infections and neurological effects. A blue baby syndrome is often caused in infants. It is due to high levels of nitrates in drinking water. Agricultural pollution also causes heavy economic losses. We should prevent agricultural pollution through planning and application of fertilizers at the correct time, planting trees and shrubs around the boundaries of farmlands, should avoid over tillage, managing the proper disposal of animal waste and anaerobic digestion of animal waste. We should use bio-fertilizers, organic fertilizers, bio-pesticides and manures to prevent agricultural pollution.

**Keywords:** Agriculture, Agricultural Pollution, Bio-Fertilizers, Contaminants, Health Problems, Organic Fertilizers, Prevention, Sources.

## INTRODUCTION

The production of food, feed, fibre and many other desired products by cultivation of certain plants is called agriculture. It is also useful for domesticated animals [Science Daily]. The agricultural practices are known as farming.

_____

* **Corresponding author K. Chitra:** Department of Botany, Bharathiar University, Coimbatore, Tamil Nadu, India; Tel: 9500287222; E-mail: drkchitraa@gmail.com

**J. Senthil Kumar, P. Ponmurugan & A. Vinothkanna (Eds.)**

Agriculture is a necessary part for the survival of human beings. Agriculture is a natural process. It was known not to cause any harm to the environment thousand years ago. Due to the increase in population, the demand for food has increased. It leads to pollution in agriculture. Farming practices lead to contamination or degradation in the environment and ecosystem through biotic and abiotic by-products, it is called agricultural pollution. To increase the yield, farmers use synthetic fertilizers, weedicides and pesticides. Such practices are harmful for the environment andinjurious to all living organisms, especially human beings. A high level of mechanization and large resource inputs are used by farmers in intensive commercial agriculture. More ecological damages are caused due to agricultural pollution than human settlements or industry in many countries [8].

In agricultural management practices, farmers use only agricultural chemicals to increase economic efficiencies in production. They need to reduce the total production costs and higher production yield. Agricultural chemicals cause potential environmental effects. But farmers give less attention to this issue [24]. Pesticide spraying significantly reduces plant diseases. The application of fertilizers provides a variety of nutrients for the growth of the crops. It increases the yield in crops. Residues of agricultural chemicals are present in the soil, water, air, agricultural products and even in human blood and adipose tissues [4, 18].

## SOURCES OF AGRICULTURAL POLLUTION

1. Point source pollution
2. Pollutants from a single discharge point.
3. Non-point source pollution.

It is from many discharge points. Non-point source pollution includes pollutants from pesticides, fertilizers, animal manure and soil washed into streams in rainfall runoff.

## CAUSES OF AGRICULTURAL POLLUTION

1. Chemical fertilizers
2. Chemical pesticides
3. Heavy metals
4. Soil tillage
5. Soil erosion
6. Soil sedimentation
7. Silage

Soil supports life on the planet. Nutrient recycling of carbon, nitrogen, sulphur and phosphorus is done by microorganisms living in the soil. Soil health and fertility are maintained by bacteria. They are included in the major class of microorganisms present in the soil. They play a key role in the decomposition of the soil. Decomposition leads to increase water infiltration and water holding capacity of the soil. These two are essential for agriculture (Fig. **1**) [20].

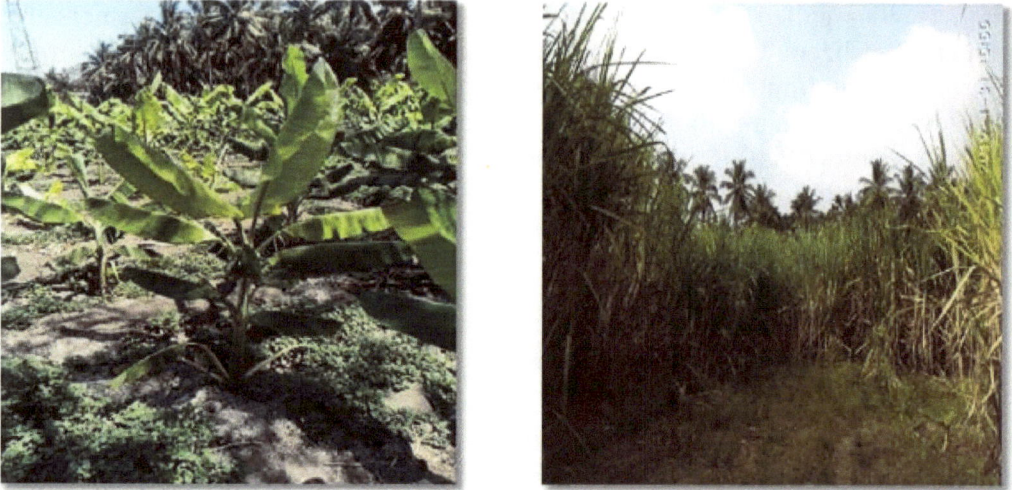

**Fig. (1).**  Agricultural areas.

## CHEMICAL FERTILIZERS

Farmers apply nutrients to agricultural fields in the form of chemical fertilizers and animal manure. They provide nitrogen and phosphorous, which are necessary for the growth of the crops. Nitrogen and phosphorus are not fully utilised by the plants. Some quantities are left to the field. Downstream water quality and air are contaminated by these chemicals. During the rainy season, these excess nutrients are washed from the farm fields and enter the waterways. Soil and groundwater are contaminated due to the leach of excess nutrients.

## EUTROPHICATION

In aquatic ecosystems, chemical nutrients are present in a higher amount. They are nitrogen and phosphorus from manure and fertilizers of agricultural chemicals. They may be washed into nearby surface waters due to rain or irrigation. It gives rise to eutrophication. Eutrophication is the dense growth of plant life and algae on the water surface. It represents the high incidences of algal blooms. Dissolved oxygen is extensively depleted in eutrophication. In the absence of oxygen, fish and other aquatic life are killed. In humans, paralytic shellfish poisoning is caused due to eutrophication; it leads to death.

Salts of potassium, phosphate, nitrate and ammonium are present in synthetic fertilizers. In agriculture, synthetic fertilizers are used in higher amount by the farmers to obtain more products, leading to soil pollution, water pollution, air pollution and some other environmental problems, because heavy metals are present in chemical fertilizers (chromium, cadmium). Radionuclides are also present in higher concentrations. Later plants accumulate these type of inorganic pollutants. Excessive use of inorganic nitrogen fertilizers contaminates the soil. Acidification is a serious threat to agricultural soils.

## CHEMICAL PESTICIDES

For pest management, farmers use pesticides. They control the effectivity of the pests on the growth and productivity of agricultural crops. Pesticides have detrimental effects when they are released to the environment. They are easily spread in the surrounding areas. In developing countries, it is a common problem. They enter into the water bodies through rainwater.

## DISTRIBUTION OF AGRICULTURAL PESTICIDES

Pesticides can easily be distributed. They are easily transported to human beings and other organisms. They are transported to other areas or into rivers by the action of wind. Water bodies act as a reservoir for pesticides. They are easily absorbed by the soil and contaminate surface and groundwater. Otherwise, the plants absorb the pesticides, which is detrimental to the growth of plants. The dissolved pesticides are consumed by living organisms. The level of pesticides is very high in water streams.

US Geological Survey took into consideration 51 major river basins and aquifer systems for contamination in agricultural areas. The pesticides were detected 97% of the time in samples from stream water in agricultural areas in USA stated by US Geological Survey [9].

In agricultural communities, indoor air pollution is caused by outside pesticide application. Pesticides are the major contributors to indoor air pollution in Japan. Pesticides were frequently detected in the air by residential environments and childcare facilities [12].

## BANNED PESTICIDES IN INDIA

1. Benomyl - fungicide
2. Carbaryl - insecticide
3. Diazinon - insecticide
4. Fenthion – insecticide
5. Linuron - herbicide

  6. Methoxy ethyl mercury chloride - fungicide
  7. Methyl Parathion - insecticide
  8. Sodium cyanide – rodenticide and insecticide
  9. Thiometon - insecticide
10. Tridemorph - fungicide

The registrations, import, manufacture, formulation, transport, and sell are prohibited and their use is completely banned in India.

1. Alachlor - herbicide
2. Trifluralin - herbicide
3. Dichlorvos - insecticide
4. Phorate - insecticide
5. Phosphamidon - insecticide
6. Triazophos - insecticide
7. Trichlorfon - insecticide

No person shall import, manufacture or formulate the above pesticides with effect from the 1$^{st}$ January, 2019. The use of the above pesticides shall be completely banned with effect from the 31$^{st}$ December, 2020. They are very toxic to aquatic organisms. They should not be used near water bodies, aquaculture or pisciculture area.

## HEAVY METALS

Heavy metal pollution threatens plant, animal and human health. It also affects the quality of the environment. They cannot be degraded or destroyed. Modern agricultural practices incorporate increased application of agrochemicals and inorganic fertilizers. The concentration of heavy metals in soil, water and air is increased due to the use of inorganic fertilizers. These heavy metals are bioaccumulative and non-biodegradable. They slowly enter plants, animals and humans through air, water and food chain [13]. Research alarms that the massive use of inorganic fertilizers is always associated with the accumulation of heavy metals. Arsenic (As), cadmium (Cd), fluorine (F), lead (Pb) and mercury (Hg) were estimated in agricultural soils worldwide [24].

## SOIL EROSION

In agriculture, soil erosion and sedimentation are greatly contributed by intensive management or inefficient land cover. The irreversible decline in fertility is due to agricultural land degradation. It is estimated that 6 million hectares of fertile land loses fertility each year due to soil erosion.

## SEDIMENTATION

Water quality is affected in various ways by the accumulation of sediments from runoff water. The transport capacity of ditches, streams, rivers and navigation channels are decreased by sedimentation. It also checks the penetration of light into water. So the aquatic biota is completely affected. The population dynamics are changed, because turbidity interferes with the feeding habits of fishes. Turbidity is the result of sedimentation in aquatic habitats. Sedimentation results in the transport and accumulation of pollutants (phosphorus and various pesticides) in aquatic habitats.

## SILAGE

Farmers should manage their silage operations correctly. In streams and rivers, silage effluent causes devastating pollution if added. Silage operations take place in summer. Massive fish killing occurs in watercourse, when the silage effluent is added as it is a significant polluting substance. During summer, the river contains low water content. It has a low dilution capacity. A small leak can cause massive damage to aquatic life.

Some ways to reduce the risk of pollution are as follows:

The most environmental-friendly way is the use of round bales for storage. After use, the silage pit should be properly sealed.

During heavy rain, slurry is not used, rather is it used summer time.

It is suggested not to spread slurry close to a watercourse, and be aware of the slope of the land nearer the watercourse.

It is also suggested not to clear slurry tanks near watercourse, stream or a river;.

Accidental runs of polluting substances are prevented by good farmyard management. Farmers should concentrate on the protection of the local environment.

## AGRICULTURAL WATER POLLUTION

Agriculture plays a major role in water pollution. In water bodies, agricultural soils discharge large quantities of agrochemicals, drug residues, organic matter, sediments and saline. Water pollution is a major issue; it affects billions of people. In China, surface water and ground water are contaminated by nitrogen. A blue baby syndrome is a fatal-borne illness in infants. This serious illness is caused by a high level of nitrates in water.

Cities and industries are vital sources of pollution in the environment. Globally recent research stated that agriculture is the leading cause of water pollution [3, 11, 16]. Discharge of large quantities of agrochemicals and nutrients [6], organic matter in agricultural soils [26], drug residues [2], and sediments are the major causative factors for pollution [15]. Salinization and alkalinisation are induced by saline drainage [10, 15, 21], microplastics [1, 19, 17] and pathogens [22, 23]. Water pollution is exacerbated by the reduction of flow in water bodies [5, 7]. Water is largely needed for agriculture and it is largely driven for crop irrigation (Fig. **2**) [14, 25].

**Fig. (2).**   Water pollution by agriculture.

## EFFECTS OF AGRICULTURAL POLLUTION

Agricultural pollution contaminates the water quality in water streams and it aso contaminates lakes, rivers, streams and oceans. Widespread contamination of waterways and ground waters is due to fertilizers, pesticides, manure, and herbicides. These contaminated waters, in turn, affect microorganisms, plants, wildlife, humans, and animals.

Agricultural chemicals contain a high level of poison. They degrade water streams and the chemical nutrients deplete dissolved oxygen, killing aquatic life and fish. Several human health problems and premature deaths are caused by agricultural pollution. Oils from farm and farm machinery also cause serious threats to humans and kill living organisms when they enter drinking water.

### Health Problems

Pesticides contain poisonous chemicals. A single episode of ingestion, and inhalation may cause harmful, and sometimes, lethal effects. The symptoms appear within 48 hours. They are acutely toxic and can cause chronic diseases. Following are some of the harmful diseases caused by pesticides in humans.

## *Acute Diseases*

1. Irritation in eye and skins
2. Allergy
3. Vomiting, diarrhoea
4. High infection in the respiratory tract
5. Headache
6. Weakness
7. Chronic diseases

## *Parkinson's Disease*

1. Asthma
2. Anxiety, depression
3. ADHD – Attention Deficit and Hyperactivity Disorder
4. Leukaemia
5. Cancer
6. Non – Hodgkin's lymphoma

## Reduce Long-Term Agricultural Yields

In agriculture, farmers use agricultural chemicals continuously. They are not aware of the negative impact of agricultural chemicals on the environment. They retain in the soil for many years and affect all biotic and abiotic components on earth. They kill beneficial organisms too, therefore, there should be an ecological balance. Nowadays, farmers have shifted attention to the use of traditional and organic farming methods.

## Soil Pollution and Depletion of Soil Fertility

In agriculture, farmers use agricultural chemicals for higher yields and pest management. The soil chemistry is also altered due to agricultural chemicals, with loss of soil fertility. A large quantity of fertile soils are treated by agricultural chemicals (Figs. **3** and **4**).

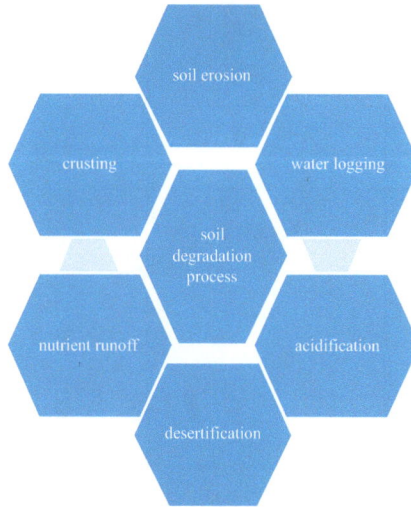

**Fig. (3).** Soil productivity – Soil degradation process.

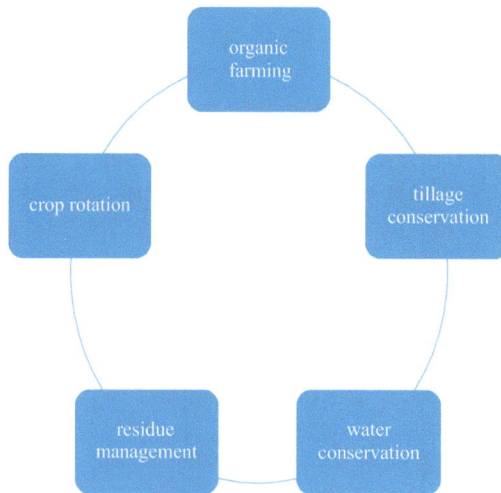

**Fig. (4).** Soil productivity - Soil conservation process.

## Air Pollution

Livestock and fertilized soils emit higher amount of nitrogen based compounds such as nitrogen oxide and ammonia to the environment. Tractors and harvesters significantly contribute to greenhouse gas emission by combusting fossil fuel. Some soil biochemical processes also naturally emit numerous greenhouse gasses. Greenhouse gases have a potential effect on the environment.

## Destroys Biodiversity

Continuous use of chemical products in agricultural production degrades and destroys the soil quality, micro-organisms, animals, plants, waters, and wildlife. It gradually alters the ecosystems which support biodiversity. Various life forms on earth are referred to as biodiversity. Biotic factors are dependent on one another such as butterflies, which have far-reaching effects on biodiversity. Agricultural chemicals deplete biodiversity in the environment.

## SOME WAYS TO REDUCE AGRICULTURAL POLLUTION

1. Farmers should follow correct farming techniques.
2. Recycling of waste before disposal should be done.
3. Household and industrial waste water should be properly recycled and disposed of.
4. Use of organic fertilizers and manures.
5. Community education and awareness for farmers must be made available.
6. Proper maintenance and disposal of the sewage system must be ensured .

## ORGANIC FARMING

Organically grown food is cultivated and processed with the use of pesticides derived from natural sources and synthetic fertilizers are never used for food products (NOSB 1995). The negative impacts of conventional farming are reduced by organic farming. Since late 1940s, organic cultivation has been practiced in the United States. The experimental garden plots are extended as larger farms and the products are formed and sold with specific organic labels. More than forty different state agencies currently certify organic food with different standards. Organic food production act was framed in the year 1990; it has a list of synthetic and non-synthetic substances. It mentions some substances that cannot be used in organic farming. Organic farming concentrates on protecting the environment and biodiversity conservation.

### Principles of Organic Agriculture

It is good for human beings because organic farming or agriculture provides chemical- and pollution-free food for humans.

It mainly concentrates on the health and well-being of plants, animals, soil, earth, and humans; it also promotes ecological, physical, and social welfare.

It is helpful in decreasing pollution and improving the quality of life.

In the living ecological system, organic farming must be modelled because it fits the environmental cycles and maintains the equilibrium of natural world.

## Benefits of Organic Farming

### *Healthy Soil*

Healthy soil definitely produces healthy food. Soil treated by harmful pesticides and chemicals cannot provide nutrients on its own. For soil management, we should follow natural cultivation practices. One teaspoon of compost-rich organic soil may contain an enormous amount of microorganisms. It contains 600 million to 1 billion helpful bacteria from 15,000 species. One teaspoon of chemically treated soil contains many bacteria.

### *Prevent Soil Erosion*

Organic farming produces healthy soil and soil erosion. Soil erosion affects the soil quality and reduces agriculture.

### *Organic Farming and Global Warming*

Healthy organic farming helps in slow climate change. Because it reduces the carbon level in the air. In organic farming, the soil serves as a carbon sink.

### *Water Conservation and Water Health*

Contaminated water is the real threat to the environment, where people and the planet both are affected by the contaminated water. Organic agriculture helps in the conservation of water. Soil amendment and mulching both help in the conservation of water in organic agriculture.

### *Support forAnimal Health and Welfare*

Animal health and welfare are preserved and supported by organic agriculture. Pest control is managed by birds and natural predators. On farm land, they live happily and control pests naturally. Organic farming provides chemical-free grazing area for animals. It keeps them naturally healthy and resistant to illness.

### *Increase in Biodiversity*

Organic agriculture helps to increase biodiversity. An increase in biodiversity is a good sign of pollution-free environment.

## CONSENT FOR PUBLICATION

Not applicable.

## CONFLICT OF INTEREST

The authors confirm that this chapter contents have no conflict of interest.

## ACKNOWLEDGEMENTS

Declared none.

## REFERENCES

[1]     Horton AA, Walton A, Spurgeon DJ, Lahive E, Svendsen C. Microplastics in freshwater and terrestrial environments: Evaluating the current understanding to identify the knowledge gaps and future research priorities. Sci Total Environ 2017; 586: 127-41.
[http://dx.doi.org/10.1016/j.scitotenv.2017.01.190] [PMID: 28169032]

[2]     Boxall ABA. New and Emerging Water Pollutants Arising From Agriculture

[3]     OECD. Paris, 2012.

[4]     Alvarez A, Saez JM, Davila Costa JS, *et al.* Actinobacteria: Current research and perspectives for bioremediation of pesticides and heavy metals. Chemosphere 2017; 166: 41-62.
[http://dx.doi.org/10.1016/j.chemosphere.2016.09.070] [PMID: 27684437]

[5]     Navarro-Ortega A, Acuña V, Bellin A, *et al.* Managing the effects of multiple stressors on aquatic ecosystems under water scarcity. The GLOBAQUA project. Sci Total Environ 2015; 503-504: 3-9.
[http://dx.doi.org/10.1016/j.scitotenv.2014.06.081] [PMID: 25005236]

[6]     Sharpley AN, Bergström L, Aronsson H, *et al.* Future agriculture with minimized phosphorus losses to waters: Research needs and direction. Ambio 2015; 44 (Suppl. 2): S163-79.
[http://dx.doi.org/10.1007/s13280-014-0612-x] [PMID: 25681975]

[7]     Kalogianni E, Vourka A, Karaouzas I, Vardakas L, Laschou S, Skoulikidis NT. Combined effects of water stress and pollution on macroinvertebrate and fish assemblages in a Mediterranean intermittent river. Sci Total Environ 2017; 603-604: 639-50.
[http://dx.doi.org/10.1016/j.scitotenv.2017.06.078] [PMID: 28667932]

[8]     FAO links water pollution to agricultural practices, News - Agriculture, August 7 2018.

[9]     Gilliom RJ. Pesticides in U.S. streams and groundwater. Environ Sci Technol 2007; 41(10): 3408-14.
[http://dx.doi.org/10.1021/es072531u] [PMID: 17547156]

[10]    Mateo-Sagasta J, Marjani S, Turral H, Eds. More people, more food, worse water? A global review of water pollution from agriculture, FAO and IWMI, Rome, 2018. This paper is a key reference for several aspects of AWP including drivers of water pollution, forms and sources of water pollution and modelling It also discusses a range of responses from policy to farm management. 2018.

[11]    Mateo-Sagasta J, Marjani S, Turral H, Eds. More people, more food, worse water? A global review of water pollution from agriculture. Rome: FAO and IWMI 2018.

[12]    Kawahara J, Horikoshi R, Yamaguchi T, Kumagai K, Yanagisawa Y. Air pollution and young children's inhalation exposure to organophosphorus pesticide in an agricultural community in Japan. Environ Int 2005; 31(8): 1123-32.
[http://dx.doi.org/10.1016/j.envint.2005.04.001] [PMID: 15979719]

[13]    Malik Z, Ahmad M, Abassi GH, Dawood M, Hussain A, Jamil M. Agrochemicals and soil microbes:

interaction for soil health.Xenobiotics in the Soil Environment: Monitoring, Toxicity and Management. Cham: Springer International Publishing 2017; pp. 139-52.
[http://dx.doi.org/10.1007/978-3-319-47744-2_11]

[14]  Bonsch M, Popp A, Biewald A, *et al.* Environmental flow provision: Implications for agricultural water and land-use at the global scale. Glob Environ Change 2015; 30: 113-32.
[http://dx.doi.org/10.1016/j.gloenvcha.2014.10.015]

[15]  Mateo-Sagasta J, Marjani S, Turral H, Eds. More people, more food, worse water? A global review of water pollution from agriculture. Rome: FAO and IWMI 2018.

[16]  OECD Policy Highlights Diffuse Pollution, Degraded Waters: Emerging Policy Solutions OECD Environment Directorate 2017.

[17]  Kay P, Hiscoe R, Moberley I, Bajic L, McKenna N, Kay P. Wastewater treatment plants as a source of microplastics in river catchments. Environ Sci Pollut Res Int 2018; 25(20): 20264-7.
[http://dx.doi.org/10.1007/s11356-018-2070-7] [PMID: 29881968]

[18]  Ridolfi AS, Alvarez GB, Rodríguez Giraul ME. Organochlorinated contaminants in general population of Argentina and other Latin American Countries.Bioremediation in Latin America Current Research and Perspectives. Springer 2014; pp. 17-40.
[http://dx.doi.org/10.1007/978-3-319-05738-5_2]

[19]  Vaughan R, Turner SD, Rose NL. Microplastics in the sediments of a UK urban lake. Environ Pollut 2017; 229: 10-8.
[http://dx.doi.org/10.1016/j.envpol.2017.05.057] [PMID: 28575711]

[20]  Deobhanj S. Scientists assess how continuous use of fertilisers affects soil bacteria. Agriculture 2018.

[21]  Kaushal SS, Likens GE, Pace ML, *et al.* Freshwater salinization syndrome on a continental scale. Proc Natl Acad Sci USA 2018; 115(4): E574-83.
[http://dx.doi.org/10.1073/pnas.1711234115] [PMID: 29311318]

[22]  Zandaryaa S, Mateo-Sagasta J. Organic matter, pathogens and emerging pollutants. In: Mateo-Sagasta J, Marjani S, Turral H, Eds. More people, more food, worse water? A global review of water pollution from agriculture. FAO & IMWI 2018; pp. 125-38.

[23]  Mateo-Sagasta J, Marjani S, Turral H, Eds. More people, more food, worse water?: A global review of water pollution from agriculture. Rome: FAO and IWMI 2018.

[24]  Udeigwe TK, Teboh JM, Eze PN, *et al.* Implications of leading crop production practices on environmental quality and human health. J Environ Manage 2015; 151: 267-79.
[http://dx.doi.org/10.1016/j.jenvman.2014.11.024] [PMID: 25585140]

[25]  Chen W, Olden JD. Designing flows to resolve human and environmental water needs in a dam-regulated river. Nat Commun 2017; 8(1): 2158.
[http://dx.doi.org/10.1038/s41467-017-02226-4] [PMID: 29255194]

[26]  Wen Y, Schoups G, van de Giesen N. Organic pollution of rivers: Combined threats of urbanization, livestock farming and global climate change. Sci Rep 2017; 7: 43289.
[http://dx.doi.org/10.1038/srep43289] [PMID: 28230079]

CHAPTER 5

# Industrial Effluent Pollution-Impact of Papermill Effluent Irrigation and Solid Waste Application on Cultivated Soil

**M. Suguna Devakumari**[*]

*Department of Agriculture, Karunya Institute of Technology and Sciences (KITS), Coimbatore – 641 114, Tamil Nadu, India*

**Abstract:** The paper mills are generating significant quantities of solid wastes and effluent. The scientific ways and means of recycling these solid wastes and effluent in an integrated, eco-friendly manner for agricultural purposes have been the main objective of the present investigation. In brief, in the present investigation, the possible options and potential for agro-cycling of the treated paper mill effluent coupled with the solid byproducts of the industry have been evaluated with special reference to Sugarcane [CO 86032]. The pH and EC of the soil were increased by 0.1 and 0.12 units due to continuous effluent irrigation. The organic carbon content and exchangeable sodium content also increased due to effluent irrigation and solid waste amendments.

**Keywords:** Biosludge, Effluent, Flyash, Pressmud.

## INTRODUCTION

Historically, the discharges from pulp and paper production have long been recognized as a significant source of pollution. Early pollution control initiatives, concentrated upon the loading of suspended solids and biological oxygen demand to the receiving waters [1]. BOD control is still problematic in areas where discharges are made to restricted watercourses or enclosed coastal areas. The pulp and paper industry is one of the largest consumers of water. Nearly 80% of freshwater used in the pulp and paper mill is discharged as effluent, containing organic and inorganic pollutants requiring treatment and disposal. The treated paper mill effluent's application to cultivated lands is considered to be an innovative approach for its disposal. By this, the effluent is not only kept out of the surface waters, but is also recycled where pollutants become the nutrients for plant growth.

[*] **Corresponding author M. Suguna Devakumari:** Department of Agriculture, Karunya Institute of Technology and Sciences (KITS), Coimbatore – 641 114, Tamil Nadu, India; Tel: 9841665854; E-mail: sugisathish@yahoo.com

**J. Senthil Kumar, P. Ponmurugan & A. Vinothkanna (Eds.)**

But continuous irrigation with industrial effluents alters the quality parameters of the receiving land, which has to be considered seriously. Hence, detailed investigations were undertaken to assess the efficiency of utilizing paper mill effluent for crop cultivation.

## REVIEW OF LITERATURE

Effluent irrigation has been practiced for centuries throughout the world. Land application of wastewater had been preferred as an alternative for its disposal since the soil is believed to have the capacity for decomposing the wastes and pollutants where organic materials are stabilized by the activities of soil microbes. The removal of different constituents is generally accomplished by physical, chemical and microbiological interaction with the soil matrix and the cover crop or plant uptake [2].

The pulp and paper mills have attempted to dispose of the lignin-rich coloured effluent through waterways on lands. The soil has the capacity to retain the lignin and the coloring matter. No appreciable decrease in hydraulic properties was noticed when the soil was irrigated with papermill wastewater for over 3 years [3]. Nila Rekha [4] reported that the soil organic matter content increased due to continuous effluent irrigation. A higher amount of organic carbon was found in the soil irrigated with effluent for over three years [5, 6].

Irrigation with undiluted paper and pulp industry effluent had led to increased soil pH [7 - 9]. Phukan and Bhattacharyya [10] also reported that the papermill effluent irrigation turned the soil pH towards alkalinity. The increase in soil EC was observed in all the seasons due to effluent irrigation and it has been attributed to the addition of a considerable quantity of dissolved salts through the effluent [11]. According to Alfred [8], the increase in K content of the soil under effluent irrigation was due to the presence of K in the effluent.

There has been a considerable increase in CEC and exchangeable cations in soils irrigated with effluent for 15 consecutive years [3]. Long term effluent irrigation increased the levels of sodium [12]. It was reported that none of the available micronutrients in the soil had reached the toxic levels due to continuous effluent irrigation [3].

## MATERIALS AND METHODS

The present investigation was carried out at the Department of Environmental Sciences, Tamil Nadu Agricultural University, Coimbatore, and in the farmer's fields at Pappampalayam, Erode. The field experiment was laid out in split-plot design with three replications.

## Treatment Details

1. Main Plot Treatments – Irrigation Sources

$I_1$ – Well water irrigation

$I_2$ – Treated effluent irrigation

2. Subplot Treatments – Solid Wastes

$T_1$ – Control (100% NPK)

$T_2$ – Biosludge @ 12.5 t ha$^{-1}$ 75% NPK $T_3$ – Pressmud @ 12.5 t ha$^{-1}$ + 75% NPK

$T_4$ – Flyash @ 20 t ha$^{-1}$ + Biosludge @ 6 t ha$^{-1}$ + 75% NPK

The sugarcane variety CO 86032 was used for this study. It is a cross selection between CO 62198 and COC 671. It is a high yielding and high sucrose-containing mid-late variety with an attractive green canopy. The canes are medium thick and reddish-pink with prominent ivory marks and purple leaf sheath. The variety is resistant to smut. Immediately after planting, the first irrigation was made possible, followed by life irrigation on the 3$^{rd}$ and 7$^{th}$ day. Then irrigation was performed once in 8 to 10 days up to 10 months and thereafter at 10 to 12 days interval up to harvest.

The initial soil sample was collected from the experimental field as per the standard procedure and analyzed for various physicochemical properties (Table 1). Soil samples were also collected at different stages of crop growth [30, 60, 90, 170, 240, 300 days after planting and at harvest] and analyzed after processing. The collected samples were air-dried, powdered with a wooden mallet and sieved through a 2 mm sieve.

**Table 1. Initial characteristics of Sugarcane experimental field.**

| S.No | Parameters | Value |
|------|------------|-------|
| 1. | pH | 7.52 |
| 2. | EC [dS m$^{-1}$] | 0.80 |
| 3. | Total N[%] | 0.32 |
| 4. | Total P [%] | 0.41 |
| 5. | Soil Available N [kg ha$^{-1}$] | 184 |
| 6. | Soil Available P [kg ha$^{-1}$] | 10.4 |
| 7. | Organic Carbon [%] | 0.48 |

*(Table 1) cont.....*

| S.No | Parameters | Value |
|------|------------|-------|
| 8. | Exchangeable Ca [ c mol (p$^+$) kg$^{-1}$] | 7.58 |
| 9. | Exchangeable Mg [ c mol (p$^+$) kg$^{-1}$] | 3.30 |
| 10. | Exchangeable Na [ c mol (p$^+$) kg$^{-1}$] | 1.83 |
| 11. | Exchangeable K [ c mol (p$^+$) kg$^{-1}$] | 1.12 |

## RESULT AND DISCUSSION

### Characteristics of The Paper Mill Effluent

The color of the effluent was light brown 150 color units were recorded. The pH of the effluent ranged from 7.1 to 7.6, whereas the EC ranged from 0.9 to1.3 dSm$^{-1}$. The TDS and TSS of the effluent ranged from 680 to 710 mgL$^{-1}$ and 20 to 30 mgL$^{-1}$, respectively. The BOD ranged from 10 to 14 mgL$^{-1}$. The range of Ca, Mg, Na and K contents was 196-216, 90-146, 123-137 and 18-19 mgL$^{-1}$, respectively, with soluble sodium percent being 19-24 mgL$^{-1}$. The ammoniacal nitrogen was between 28 –30 mgL$^{-1}$.

The highest mean pH of 7.85 and 7.75 was observed in effluent irrigation -I$_2$ and well water irrigation -I$_1$, respectively. There was a marginal progressive increase in pH with the advancement of crop growth. In the present study, the increase in pH of the soil in the treated effluent receiving plots to the tune of only about 0.10 units might be due to continuous irrigation with the effluent, which is slightly alkaline in nature and increased the salt accumulation in the soil (Table **2**). This was in line with the previous findings [13, 14].

Table 2. Soil pH as influenced by effluent irrigation and solid wastes application under sugarcane crop.

| Treatments | 30 D | 60 D | 90 D | 170 D | 240 D | 300 D | Harvest | Mean |
|------------|------|------|------|-------|-------|-------|---------|------|
| I$_1$T$_1$ | 7.55 | 7.64 | 7.67 | 7.71 | 7.73 | 7.75 | 7.75 | **7.69** |
| T$_2$ | 7.63 | 7.71 | 7.75 | 7.78 | 7.80 | 7.85 | 7.86 | **7.77** |
| T$_3$ | 7.59 | 7.64 | 7.69 | 7.73 | 7.77 | 7.79 | 7.80 | **7.72** |
| T$_4$ | 7.68 | 7.75 | 7.80 | 7.84 | 7.87 | 7.89 | 7.91 | **7.82** |
| **Mean** | **7.61** | **7.69** | **7.73** | **7.77** | **7.79** | **7.82** | **7.83** | 7.75 |
| I$_2$T$_1$ | 7.63 | 7.74 | 7.78 | 7.82 | 7.85 | 7.88 | 7.88 | **7.80** |
| T$_2$ | 7.73 | 7.81 | 7.86 | 7.89 | 7.93 | 7.95 | 7.96 | **7.88** |
| T$_3$ | 7.70 | 7.79 | 7.83 | 7.85 | 7.87 | 7.89 | 7.90 | **7.83** |
| T$_4$ | 7.76 | 7.85 | 7.89 | 7.92 | 7.95 | 7.99 | 8.00 | **7.91** |
| **Mean** | **7.71** | **7.80** | **7.84** | **7.87** | **7.90** | **7.93** | **7.94** | **7.85** |

*(Table 2) cont.....*

|  | I | T | I x T | D | DxI | DxT |
|---|---|---|---|---|---|---|
| **SEd** | 0.03 | NS | NS | 0.00 | NS | NS |
| **CD [0.05]** | 0.07 | NS | NS | 0.00 | NS | NS |

$I_1$ – Well water irrigation
$T_1$ – Control [100% NPK]
$T_2$ – Bio sludge 12.5 t ha$^{-1}$ + 75% NPK
$T_3$ – Press mud 12.5 t ha$^{-1}$ + 75% NPK
$T_4$ – Fly ash 20 t ha$^{-1}$ + Bio sludge 6 t ha$^{-1}$ + 75% NPK
$I_2$ – Effluent irrigation

The EC of the soil increased by 0.12 units under effluent irrigation at the end of the crop growth period (Table **3**). Effluent irrigation contributed to 0.15 units higher soil EC over well water irrigation. Due to the solid waste application, the EC increased continuously in the treated effluent irrigated plots. The highest EC was recorded in treatment $I_1T_4$ receiving Fly ash 20 t ha$^{-1}$ + Bio sludge 6 t ha$^{-1}$ + 75% NPK under effluent irrigation wherein an increase of 0.42 units over that of control (100% NPK) under well water irrigation was observed. The higher EC in the treated effluent and fly ash + Bio sludge receiving treatments might be due to salt accumulation contributed by the solid wastes and effluent. Several other workers also inferred the same [15, 16].

**Table 3. Soil EC [dS m$^{-1}$] as influenced by effluent irrigation and solid wastes application under sugarcane crop.**

| Treatments | 30 D | 60 D | 90 D | 170 D | 240 D | 300 D | Harvest | Mean |
|---|---|---|---|---|---|---|---|---|
| $I_1T_1$ | 0.91 | 0.93 | 0.95 | 0.96 | 0.98 | 1.00 | 1.00 | **0.96** |
| $T_2$ | 1.00 | 1.01 | 1.03 | 1.04 | 1.05 | 1.06 | 1.08 | **1.04** |
| $T_3$ | 1.01 | 1.03 | 1.05 | 1.06 | 1.08 | 1.09 | 1.11 | **1.06** |
| $T_4$ | 1.22 | 1.24 | 1.26 | 1.28 | 1.29 | 1.32 | 1.35 | **1.28** |
| **Mean** | **1.04** | **1.05** | **1.07** | **1.09** | **1.10** | **1.12** | **1.14** | **1.09** |
| $I_2T_1$ | 1.03 | 1.05 | 1.06 | 1.08 | 1.10 | 1.13 | 1.14 | **1.08** |
| $T_2$ | 1.15 | 1.17 | 1.19 | 1.21 | 1.22 | 1.25 | 1.27 | **1.21** |
| $T_3$ | 1.24 | 1.25 | 1.26 | 1.29 | 1.32 | 1.33 | 1.33 | **1.29** |
| $T_4$ | 1.31 | 1.33 | 1.37 | 1.38 | 1.40 | 1.42 | 1.44 | **1.38** |
| **Mean** | **1.18** | **1.20** | **1.22** | **1.24** | **1.26** | **1.28** | **1.30** | **1.24** |

(Table 3) cont.....

| | I | T | I x T | D | DxI | DxT |
|---|---|---|---|---|---|---|
| **SEd** | 0.00 | 0.01 | NS | 0.00 | NS | NS |
| **CD [0.05]** | 0.01 | 0.01 | NS | 0.00 | NS | NS |

$I_1$ – Well water irrigation
$T_1$ – Control [100% NPK]
$T_2$ – Bio sludge 12.5 t ha$^{-1}$ + 75% NPK
$T_3$ – Press mud 12.5 t ha$^{-1}$ + 75% NPK
$T_4$ – Fly ash 20 t ha$^{-1}$ + Bio sludge 6 t ha$^{-1}$ + 75% NPK
$I_2$ – Effluent irrigation

The increase was 6.8 percent in the effluent irrigated plots compared to well water irrigated plots. The increase in the organic carbon content of the treated effluent irrigated soil might be due to the addition of suspended and dissolved solids present in the effluent. The organic carbon content revealed significant differences and the highest content was recorded in the treatment receiving the solid by-product Fly ash + Bio sludge both in an effluent irrigated and well water irrigated plots which were 45% and 40% over $T_1$, respectively (Table **4**). This was in agreement with the findings of several workers [13, 17].

Table 4. Soil organic carbon [%] as influenced by effluent irrigation and solid wastes application under sugarcane crop.

| Treatments | 30 D | 60 D | 90 D | 170 D | 240 D | 300 D | Harvest | Mean |
|---|---|---|---|---|---|---|---|---|
| $I_1T_1$ | 0.51 | 0.56 | 0.59 | 0.61 | 0.63 | 0.65 | 0.65 | **0.60** |
| $T_2$ | 0.60 | 0.65 | 0.68 | 0.74 | 0.78 | 0.81 | 0.83 | **0.73** |
| $T_3$ | 0.65 | 0.70 | 0.74 | 0.77 | 0.82 | 0.85 | 0.87 | **0.77** |
| $T_4$ | 0.71 | 0.77 | 0.79 | 0.85 | 0.89 | 0.92 | 0.94 | **0.84** |
| **Mean** | **0.62** | **0.67** | **0.70** | **0.74** | **0.78** | **0.81** | **0.82** | **0.73** |
| $I_2T_1$ | 0.56 | 0.61 | 0.63 | 0.68 | 0.71 | 0.73 | 0.75 | **0.67** |
| $T_2$ | 0.63 | 0.69 | 0.74 | 0.77 | 0.82 | 0.85 | 0.87 | **0.77** |
| $T_3$ | 0.67 | 0.74 | 0.78 | 0.83 | 0.88 | 0.90 | 0.91 | **0.82** |
| $T_4$ | 0.74 | 0.79 | 0.84 | 0.88 | 0.92 | 0.94 | 0.95 | **0.87** |
| **Mean** | **0.65** | **0.71** | **0.75** | **0.79** | **0.83** | **0.86** | **0.87** | **0.78** |

(Table 4) cont.....

|  | I | T | I x T | D | DxI | DxT |
|---|---|---|---|---|---|---|
| **SEd** | 0.00 | 0.00 | 0.01 | 0.00 | NS | 0.01 |
| **CD [0.05]** | 0.01 | 0.01 | 0.01 | 0.00 | NS | 0.02 |

$I_1$ – Well water irrigation
$I_2$ – Effluent irrigation
$T_1$ – Control [100% NPK]
$T_2$ – Bio sludge 12.5 t ha$^{-1}$ + 75% NPK
$T_3$ – Press mud 12.5 t ha$^{-1}$ + 75% NPK
$T_4$ – Fly ash 20 t ha$^{-1}$ + Bio sludge 6 t ha$^{-1}$ + 75% NPK

There were significant increases in exchangeable sodium and potassium contents of the soil under-treated effluent irrigation. The exchangeable cationic concentrations were the highest in fly ash applied plots (Tables **5** and **6**). The results were in accordance with the findings of Bharti Bhaisari [18]. The exchangeable cations increased with the duration of effluent irrigation. This was in agreement with the findings of Matli Srinivasachari [19]. The higher exchangeable cations in soils irrigated with effluent could be ascribed to the accumulation of cations present in the effluent. This has been inferred by Pushpavalli [20].

Table 5. Soil exchangeable sodium [c mol (p$^+$) kg$^{-1}$] as influenced by effluent irrigation and solid wastes application under sugarcane crop.

| Treatments | 30 D | 60 D | 90 D | 170 D | 240 D | 300 D | Harvest | Mean |
|---|---|---|---|---|---|---|---|---|
| $I_1T_1$ | 1.79 | 2.02 | 2.18 | 2.21 | 2.23 | 2.26 | 2.27 | **2.14** |
| $T_2$ | 1.94 | 2.27 | 2.48 | 2.59 | 2.65 | 2.67 | 2.71 | **2.47** |
| $T_3$ | 1.86 | 2.13 | 2.31 | 2.47 | 2.53 | 2.55 | 2.59 | **2.35** |
| $T_4$ | 1.92 | 2.24 | 2.43 | 2.56 | 2.62 | 2.63 | 2.68 | **2.44** |
| **Mean** | **1.88** | **2.17** | **2.35** | **2.46** | **2.51** | **2.53** | **2.56** | **2.35** |
| $I_2T_1$ | 1.84 | 2.06 | 2.23 | 2.37 | 2.41 | 2.44 | 2.45 | **2.26** |
| $T_2$ | 1.97 | 2.29 | 2.51 | 2.64 | 2.70 | 2.71 | 2.73 | **2.51** |
| $T_3$ | 1.89 | 2.17 | 2.34 | 2.43 | 2.52 | 2.54 | 2.55 | **2.35** |
| $T_4$ | 1.93 | 2.27 | 2.45 | 2.59 | 2.64 | 2.65 | 2.66 | **2.46** |
| **Mean** | **1.91** | **2.20** | **2.38** | **2.51** | **2.57** | **2.59** | **2.60** | **2.39** |

| | I | T | I x T | D | DxI | DxT |
|---|---|---|---|---|---|---|
| **SEd** | 0.01 | 0.02 | 0.02 | 0.00 | NS | 0.03 |
| **CD (0.05)** | 0.02 | 0.03 | 0.04 | 0.01 | NS | 0.07 |

$I_1$ – Well water irrigation
$I_2$ – Effluent irrigation
$T_1$ – Control [100% NPK]
$T_2$ – Bio sludge 12.5 t ha $^{-1}$ + 75% NPK
$T_3$ – Press mud 12.5 t ha $^{-1}$ + 75% NPK
$T_4$ – Fly ash 20 t ha $^{-1}$ + Bio sludge 6 t ha $^{-1}$ + 75% NPK

**Table 6. Soil exchangeable potassium [c mol (p⁺) kg⁻¹] as influenced by effluent irrigation and solid wastes application under sugarcane crop.**

| Treatments | 30 D | 60 D | 90 D | 170 D | 240 D | 300 D | Harvest | Mean |
|---|---|---|---|---|---|---|---|---|
| $I_1T_1$ | 1.14 | 1.12 | 1.10 | 1.07 | 1.06 | 1.05 | 1.00 | **1.08** |
| $T_2$ | 1.39 | 1.38 | 1.34 | 1.32 | 1.31 | 1.30 | 1.28 | **1.33** |
| $T_3$ | 1.31 | 1.29 | 1.25 | 1.23 | 1.20 | 1.20 | 1.18 | **1.24** |
| $T_4$ | 1.45 | 1.44 | 1.41 | 1.39 | 1.36 | 1.33 | 1.30 | **1.38** |
| **Mean** | **1.32** | **1.31** | **1.28** | **1.25** | **1.23** | **1.22** | **1.19** | 1.26 |
| $I_2T_1$ | 1.16 | 1.14 | 1.15 | 1.11 | 1.09 | 1.07 | 1.04 | **1.11** |
| $T_2$ | 1.46 | 1.43 | 1.40 | 1.37 | 1.35 | 1.33 | 1.31 | **1.38** |
| $T_3$ | 1.33 | 1.31 | 1.27 | 1.26 | 1.23 | 1.21 | 1.18 | **1.26** |
| $T_4$ | 1.50 | 1.47 | 1.43 | 1.40 | 1.37 | 1.34 | 1.33 | **1.41** |
| **Mean** | **1.36** | **1.34** | **1.31** | **1.29** | **1.26** | **1.24** | **1.22** | **1.29** |

| | I | T | I x T | D | DxI | DxT |
|---|---|---|---|---|---|---|
| **SEd** | 0.01 | 0.01 | NS | 0.00 | NS | NS |
| **CD [0.05]** | 0.01 | 0.02 | NS | 0.00 | NS | NS |

$I_1$ – Well water irrigation
$I_2$ – Effluent irrigation
$T_1$ – Control [100% NPK]
$T_2$ – Bio sludge 12.5 t ha $^{-1}$ + 75% NPK
$T_3$ – Press mud 12.5 t ha $^{-1}$ + 75% NPK
$T_4$ – Fly ash 20 t ha $^{-1}$ + Bio sludge 6 t ha $^{-1}$ + 75% NPK

## CONCLUSION

The increase in pH of the soil in the treated effluent receiving plots was to the tune of only about 0.10 units during one year of the crop season. The EC of the soil was increased by 0.12 units under effluent irrigation after the period of crop growth. Effluent irrigation contributed to 0.15 units higher soil EC over well water irrigation. The increase in organic carbon was 6.8 percent in the effluent irrigated plots compared to well water irrigated plots. The highest organic carbon content was recorded in the treatment receiving the solid by-products Fly ash + Bio sludge both under effluent irrigated and well water irrigated plots. The soil exchangeable sodium and potassium also considerably increased under effluent irrigation. Therefore, it can be concluded that irrigation with effluent for a long period alters the physicochemical properties of the soil, hence alternate irrigation with fresh water is recommended to leach out the excess salts.

## CONSENT FOR PUBLICATION

Not applicable.

## CONFLICT OF INTEREST

The authors confirm that this chapter contents have no conflict of interest.

## ACKNOWLEDGEMENTS

Declared none.

## REFERENCES

[1]     Waldmeyer T. The treatment of papermill wastes. In: Isaac PCG, Ed. The treatment of trade wastewaters and the prevention of river pollution. University of Durham, King's College 1957; pp. 215-26.

[2]     Young JC, Mc Dermott GN, Jenkins D. Alterations in the BOD procedures for the 15th edition of standard methods for the examination of water and waste water. J Wat Pollut Contr Fed 1981; 53: 1253-9.

[3]     Palaniswami C, Sree Ramulu US. Effects of continuous irrigation with paper factory effluent on soil properties. J Indian Soc Soil Sci 1994; 42: 139-40.

[4]     Nila Rekha P, Ambujam NK, Ramachandran S. Irrigation – solution to industrial pollution. Abst 10th Congress of APD – IAHR. Langkawi Island, Malaysia 1996; pp. 45-7.

[5]     Palaniswami C. Studies on the effect of continuous irrigation with paper factory effluent on soil properties and on sugarcane nursery and main crop and development of techniques for clarification of effluent. MSc [Ag] thesis. Coimbatore: Tamil Nadu Agricultural University 1990.

[6]     Reddy MR, Jivendran S, Jain SC. Paper mill effluent for sugarcane irrigation. IAWPC Technol Annu 1981; 8: 129-46.

[7]     Datta M, Gupta RK. Utilization of pressmud as amendment of acid soils in Nagaland. J Indian Soc Soil Sci 1983; 31: 511-6.

[8]     Alfred RS. Impact of paper and pulp factory effluent on soil, water and plant eco-system. PhD [ENS] Dissertation. TNAU, Coimbtore 1998.

[9]     Dhevagi P. Studies on the impact of paper mill effluent on agro-ecosystem. PhD [ENS] Dissertation, TNAU, Coimbatore 1996.

[10]    Phukan S, Bhattacharyya KG. Modification of soil quality near a pulp and paper mill. Water Air Soil

Pollut 2003; 146: 319-33.
[http://dx.doi.org/10.1023/A:1023902222630]

[11]   Matli S, Dhakshinamoorthy M, Arunachalam G. Accumulation and availability of Zn, Cu, Mn and Fe in soils polluted with paper mill wastewater. Madras Agric J 2000; 87: 237-40.

[12]   Ponniah C. Impact of pulp and paper mill effluent in fodder grass soil ecosystem. PhD [Env Science] Thesis. Coimbatore: Tamil Nadu Agrl. University 1997.

[13]   Rajannan G, Oblisami G. Effect of paper factory effluents on the soil and crop plants. Indian J Environ Health 1979; 21: 120-30.

[14]   Vasconcelos E, Cabral F. Use and environmental implications of pulp-mill sludge as an organic fertilizer. Environ Pollut 1993; 80(2): 159-62.
[http://dx.doi.org/10.1016/0269-7491(93)90142-B] [PMID: 15091859]

[15]   Sandana KMC. Studies on the effect of liquid and solid wastes from industries and growth of certain crops. MSc [Ag] Thesis. Coimbatore: Tamil Nadu Agricultural University 1995.

[16]   Subrahmanyam PVR, Juwarkar AS, Sundaresan BB. Utilization of pulp and papermill wastewater for crop irrigation. Asian Chemical Conference on priorities in chemistry in Development of Asia. 26-31.

[17]   Somashekar RT, Gowda MTG, Shettigar SLB, Srinath KP. Effect of industrial effluents on crop plants. Indian J Environ Health 1984; 26: 136-46.

[18]   Bharti Bhaisare DB, Matte WP, Badole AD, Pillewan S. Effect of flyash on physico-chemical properties of vertisol and yield of green gram. J Soils Crops 1996; 8: 255-7.

[19]   Matli S, Dhakshinamoorthy M, Arunachalam G. Effect of paper factory effluent on soil available macronutrients and yield of rice. Madras Agric J 1998; 85: 564-6.

[20]   Pushapvalli R. Studies on the characterization of pulp and paper mill effluent and its effect on soil profile characteristics and on germination, yield and juice quality of sugarcane [Var. CO 63004 and COC 671]. MSc [Ag] thesis. Coimbatore: Tamil Nadu Agricultural University 1990.

# CHAPTER 6

# Genetically Modified Plants and Its Effect on Environment

**J. Beslin Joshi**[*]

*Department of Plant Biotechnology, Centre for Plant Molecular Biology and Biotechnology, Coimbatore – 641 003, Tamil Nadu, India*

**Abstract:** Population increase, shrinkage of arable land and available resources have led to the development of genetically modified plants for improving food quality and quantity. Cultivation of genetically modified plants has been reported to improve the socio- economic condition of farmers around the globe and their cultivated area has increased constantly. Important food, and feed crops are genetically engineered for improving the nutritional quality, cooking/edible quality, herbicide tolerance, shelf life, pest and disease resistance. Besides several advantages, the impact of genetically modified plants on human health, environment, social and political conditions needs to be addressed. Educating common public towards genetically modified plants and their safe use; framing proper policies and strict regulations towards transgene escape, loss of diversity, non-target effects will empower the society to relish the benefits of genetically modified plants.

**Keywords:** Genetically Modified Plants, Genome Engineering, GM Trait, Impact, Regulations.

## INTRODUCTION

Genetically modified plants are generated by genome modification/engineering through biotechnological means. Succeeding the invention of molecular scissors and ligase enzymes, the recombinant DNA technology has proved fruitful. Though DNA modifying enzymes were discovered during 1970s, a genetically modified plant (tobacco) containing an antibiotic-resistant gene was first produced in the year 1983. Since then, recombinant DNA technology has made inconceivable modifications in the genome of an organism for the well-being of mankind and has conquered almost all the fields of science. Among the different

---

[*] **Corresponding author J. Beslin Joshi:** Department of Plant Biotechnology, Centre for Plant Molecular Biology and Biotechnology, Coimbatore – 641 003, Tamil Nadu, India; Tel: 9486738122; E-mail: beslinjoshi@gmail.com

**J. Senthil Kumar, P. Ponmurugan & A. Vinothkanna (Eds.)**

genetically modified organisms, genetically modified crops are an important subject of discussion, since we directly rely on plants for nutrition that enter the food chain and in turn, affect the biodiversity of earth.

## IS GM CROPS A NECESSITY?

Around the globe, human population is constantly growing at the rate of 83 million people every year and in 2050, the world population is expected to be 9.8 billion [1]. Major population growth, hunger and malnutrition have been reported in the developing countries of the world [2]. Urbanisation had resulted in soil degradation, depletion of water resources and climate change. As a result, the area under cultivation has reduced drastically and the challenge of the hour is to produce good quality food from the limited arable land with the available resources. Earlier in 1960s, the widespread famine was overcome by developing high yielding, disease-resistant wheat and rice cultivars by Norman Borlaug and Henry Beachell, respectively through a breeding program [3, 4]. In the present context, relying completely on conventional breeding approaches for increasing crop productivity is impossible, since it takes several years to develop a high yielding variety and moreover, the genetic variations in the gene pool of sexually compatible crops have become limited. Under such circumstances, genetic engineering offers the advantage of gene pool utilisation across the kingdom and stacking of multiple genes for various traits within a shorter time span. Till date, genes for nutritional improvement, higher yield, input use efficiency, drought tolerance, salinity tolerance, herbicide resistance, pest resistance, disease resistance, delayed senescence and flower colour were introduced from other organisms into plants successfully [5].

## STATUS OF GM CROPS IN THE WORLD

Around the globe, GM crops were cultivated in 189.8 million hectares during 2017. The area under GM crops has been constantly increasing and in 2017, the raise was about 3% compared to the previous year [6]. The wide adaptability of GM crops by farmers was due to their enormous benefits like good quality produce, higher yield, low pesticide/herbicide use leading to higher returns and in turn, improving the socio-economic conditions of farmers [7]. Around 56% of GM crops were planted in developing countries and its adaptation has reduced poverty, malnutrition, GH gas emission and usage of herbicide, pesticide chemicals [8, 9]. Twenty-four countries in the world are growing GM crops and about 77% of the area is occupied by soybean, 80% by cotton, 32% by maize, and 30% by canola [6]. The first genetically modified crop approved for release in 1994 was Flavr Savr tomato, developed by Calgene for longer shelf-life [10]. Since then, several crops like maize, cotton, canola, rice, *etc.,* have been engineered for several traits *viz.,* nutritional improvement, herbicide tolerance, disease and insect resistance. The list of GM crops released for commercial cultivation in India is provided in Table **1**.

Table 1. List of crops and the Genetically Modified trait approved for commercialization around the world.

| S.No | Crop | Scientific Name | GM Trait |
|---|---|---|---|
| 1. | Alfalfa | *Medicago sativa* | Glyphosate herbicide tolerance Altered lignin production |
| 2. | Apple | *Malus domestica* | Non-Browning |
| 3. | Argentine Canola | *Brassica napus* | Modified oil/fatty acid; Herbicide (Glufosinate, Glyphosate) tolerance; Male sterility; Fertility restoration; Phytase production |
| 4. | Bean | *Phaseolus vulgaris* | Bean Golden Mosaic Virus (BGMV) Resistance |
| 5. | Carnation | *Dianthus caryophyllus* | Sulfonylurea herbicide tolerance Modified flower color |
| 6. | Chicory | *Cichorium intybus* | Glufosinate herbicide tolerance Male sterility |
| 7. | Cotton | *Gossypium hirsutum L.* | Herbicide (Glufosinate, Glyphosate) tolerance; Lepidopteran insect resistance |
| 8. | Cowpea | *Vigna unguiculata* | Lepidopteran insect resistance |
| 9. | Creeping Bent grass | *Agrostis stolonifera* | Glyphosate herbicide tolerance |
| 10. | Eggplant | *Solanum melongena* | Lepidopteran insect resistance |
| 11. | Eucalyptus | *Eucalyptus sp.* | Volumetric Wood Increase |
| 12. | Flax | *Linum usitatissimum L.* | Sulfonylurea herbicide tolerance |
| 13. | Maize | *Zea mays L.* | Multiple insect resistance Mannose metabolism Herbicide (Glufosinate, Glyphosate) tolerance Coleopteran insect resistance |
| 14. | Melon | *Cucumis melo* | Delayed ripening/senescence |
| 15. | Papaya | *Carica papaya* | Papaya Ring Spot Virus (PRSV) resistance |
| 16. | Petunia | *Petunia hybrida* | Modified flower color |
| 17. | Plum | *Prunus domestica* | Viral disease resistance |
| 18. | Polish canola | *Brassica rapa* | Herbicide (Glufosinate, Glyphosate) tolerance |
| 19. | Poplar | *Populus sp.* | Lepidopteran insect resistance [REMOVED HYPERLINK FIELD]Multiple insect resistance |
| 20. | Potato | *Solanum tuberosum L.* | Modified starch/carbohydrate Lowered Free Asparagine Lowered Reducing Sugars Coleopteran insect resistance Viral disease resistance Glyphosate herbicide tolerance Foliar Late Blight Resistance |

| S.No | Crop | Scientific Name | GM Trait |
|---|---|---|---|
| 21. | Rice | *Oryza sativa L.* | Lepidopteran insect resistance Glufosinate herbicide tolerance<br>Mannose metabolism; Enhanced Provitamin A Content |
| 22. | Rose | *Rosa hybrida* | Modified flower color |
| 23. | Safflower | *Carthamus tinctorius L.* | Modified oil/fatty acid |
| 24. | Soybean | *Glycine max L.* | Lepidopteran insect resistance<br>Herbicide (Glufosinate, Glyphosate, Mesotrione and 2,4-D) tolerance |
| 25. | Squash | *Cucurbita pepo* | Watermelon mosaic potyvirus 2 resistance |
| 26. | Sugar Beet | *Beta vulgaris* | Glyphosate herbicide tolerance |
| 27. | Sugarcane | *Saccharum sp* | Drought stress tolerance Lepidopteran insect resistance |
| 28. | Sweet pepper | *Capsicum annuum* | Cucumber mosaic cucumovirus (CMV) resistance |
| 29. | Tobacco | *Nicotiana tabacum L.* | Nicotine reduction; Oxynil herbicide tolerance |
| 30. | Tomato | *Lycopersicon esculentum* | Delayed fruit softening<br>Cucumber mosaic cucumovirus (CMV) resistance |
| 31. | Wheat | *Triticum aestivum* | Glyphosate herbicide tolerance |

Source: https://www.isaaa.org/gmapprovaldatabase/default.asp

## IMPACT OF GM CROPS

The impact of GM crops on human health, environment, social, economical aspects is discussed below.

### Human Health

#### *Allergenicity/Toxicity*

Few proteins in the natural food cause allergic reaction to humans when consumed. Since proteins are engineered in GM crops, public are concerned about the allergenicity of the engineered product. Proteins engineered through genetic modification are not allergic unless an allergic protein is introduced. Hence extensive food safety studies on the engineered product need to be carried out to prevent introduction of a new allergen and if it is expected to have toxicity, it should be abandoned. Transfer of Brazil nut gene into soybean [11] and bean crop engineered for high cysteine and methionine contents was abandoned, since it was reported to have allergenicity [12]. Except these two reports, no allergens were reported in GM crops. Besides allergic tests in animal models, the GM foods have also been tested against the sera of known allergic patients to evaluate whether the engineered gene codes for a similar type of allergens. No allergic reactions due to pollen from GM crops have been reported to date [13].

## Horiozontal Gene Transfer

Incomplete digestion of GM food over consumption will lead to the presence of rDNA in the gastrointestinal tract and can be absorbed by the intestinal cell lining or gut microflora. In general, the DNA inside a cell will be degraded, but the bacteria have the tendency to absorb DNA from the environment [14]. Transfer of selectable marker genes like antibiotics may lead to the development of antibiotic-resistant bacteria. Hence prior to the approval of GM crops, their safety in terms of human health and the environment should be examined carefully. The antibiotic resistance gene pollution source tracking systems are under development based on several algorithms [15 - 17]. Recently, Li and his co-workers showed that SourceTracker, along with a proper metagenomics approach will help in identifying antimicrobial resistant gene pollution source, so that effective control strategies can be designed [18].

## Environment

### Biodiversity Loss/Genetic Erosion

Continuous cultivation of GM crops may lead to the genetic erosion of natural bioresource. When a tolerant or resistant gene is introduced, it is more prone to survive than the susceptible gene, leading to the elimination or contamination of the susceptible gene from the natural gene pool [19, 20]. Hence after the release of GM crops, the risk of biodiversity loss should be continuously monitored.

### Super Weeds

The flow of herbicide or insect resistance genes from GM crops into natural gene pool by cross-pollination may lead to the development of "Super weeds" which are difficult to control [21]. This can be prevented by physical isolation and genetic containment [22]. The transgene escape can be regulated by using male-sterile lines, a chemical inducible promoter and split genes [23]. Transfer of engineered genes through pollen from GM crops can be restricted by chloroplast-mediated transformation [24].

### Gene Flow

The transgene may escape from transgenic plants through seed or pollen. This mixing up was reported by Cban in 2015 [25], leading to unpredicted ecological change. The escape of herbicide resistant gene in *B. napus* at a distance of 200 and 400m was reported [26]. In Mexico, the contamination of native maize varieties with GM corn pollen was reported during 2001 and later, it could not be validated [27, 28]. GM creeping bentgrass was found to pollinate with wild type bent grass up to 3.8 km in Oregon [29]. Hence the pollen viability, pollen flow distance, insect flora related to the crop, their wild relatives and weeds associated should be analysed case by case.

## *Non-Target Effect*

The development of resistant crops through genetic engineering may lead to the ecological imbalance in the food cycle. The GM crops may reduce the prey/host of another organism and their food chain may be affected, leading to an imbalance in the natural ecosystem [30]. The most widely discussed case is the Monarch butterfly. The larvae fed on milkweed dusted with GM corn pollen containing *Bt* gene reduced survival [31]. Apart from *Bt* gene containing pollen, the same risk is caused by spraying *Bt* insecticide which needs further assessment [32]. Such issues/risks should be analysed before the release of GM crops [33].

## Economical

Though the use of GM crops has raised the income of farmers by reducing the input use, patenting and engineering new genes may increase the seed price so high affecting small farmers [34]. This can be regulated through policies on seed quality and seed price.

## Social

GM crops will affect the social interaction of farmers and food security in the rural community [35]. Cartagena protocol emphasises on the acceptance, transfer, and import of GM food/feed based on socioeconomic affairs [36].

## Political

The use and import of GM products are also purely based on political concerns. Political parties play a major role in forming regulations based on the factors that affectthe trade and environmental regulation. GM plants should be labelled properly and there should be a common law in regulating all countries [37].

## REGULATION OF GM CROPS

The regulation of GM crops was established in early 1990s globally and since then, several modifications have been made by many countries. In order to differentiate natural foods from GM foods, labelling is mandatory so the consumer can select their food based on their choices. Regulation of rules for GMOs is based on political, social, economical and environmental concerns. GM crop regulations in some of the GM crop cultivating countries are discussed below.

## US

US is the leading GM crop cultivation country in the globe and about 70-80% of processed food is from GM plants. The U.S. Department of Agriculture (USDA), the Environmental Protection Agency (EPA) and the Food and Drug Administration (FDA) regulate the GM crops in the USA. Field release of GM crops was approved by the USDA, EPA is involved in pest and pesticide resistance traits, and FDA regulates the crops destined for food, feed, or pharmaceuticals [38]. Though in earlier days, GM foods were accepted by US consumers, but recently voices were raised against GM foods. This has insisted GM food labelling a mandatory process, which is currently a voluntary process [39].

## EU

European Commission (EC) and the European Food Safety Authority regulate the rule for GMOs in EU. GM foods can be marketed in the European countries after proper approval. In case of objection from member states, the applicant can apply additional evaluation, and voting is carried out. Unlike the US, the policy makers in EU are anti-GM and GM labelling is compulsory in EU [40].

## China

Though GM regulation in china started earlier during 1990, safety assessment and labelling of GMOs were made compulsory in 2002 [41]. Food safety law in China has been updated and is strictly followed. Consumers in China were educated about GMOs and they could select their food based on their interest.

## Japan

In Japan, GMOs were regulated under the Cartagena protocol. The Ministry of Education, Culture, Sports, Science and Technology monitors the experimental trials related to GM crops and their impact on the environment. Biodiversity is assessed by the Ministry of Environment along with the Ministry of Agriculture, Forestry and Fisheries (MAFF). The safety of GM crops in food is assessed by the Ministry of Health, Labour and Welfare; while in feed, by MAFF [42]. In a processed food, if the GM crop content exceeds 5-10% of the total weight or is among the top 3 ingredients, then GMO labelling is mandatory on that product [43].

## India

In India, the Environment (Protection) Act 1986 introduced by the Ministry of Environment, Forest and Climate Change (MoEFCC), holds the apex rules for

GMO regulation. In 1989, the rules for manufacture, use, import, export and storage of GMOs were notified by MoEFCC and the following six authorities are obliged to regulate the process [44].

1. RDAC- Recombinant DNA Advisory Committee
2. IBSC – Institutional Biosafety committee
3. RCGM – Reginal Committee on Genetic Manipulation
4. GEAC- Genetic Engineering Appraisal Committee
5. SBCC- State Biotechnology Coordination Committee
6. DLC – District Level Committee

RDAC, constituted by DBT prepared the recombinant DNA biosafety guidelines in 1990 and was adopted for handling GMOs for conducting research including the import of an organism for research work in India. IBSC operates in each institution and its members are the organisation head, scientists, medical expert and a nominee from DBT. On receiving the application by IBSC, it was forwarded to RCGM. RCGM prepares the procedure for the regulatory process. GEAC is the apex committee and is chaired by MoEFCC senior officer and co-chaired by DBT expert. SBCC and DLC are for monitoring purposes. In 2008, ICMR prepared the guidelines for the safety assessment of GMOs based on the Codex Alimentarius commission 2003 and for safety field trial, rDNA research by DBT and MoEFCC [45].

## CONCLUSION

Owing to recent developments in biotechnology, improvement in crop health, productivity and nutrition within a short time span have been seen. In order to feed the growing population from the limited arable land and changing climatic conditions, GM crops are the best solution. Biosafety issues related to genetically modified crops like gene escape, adverse effects, non-target organisms and allergenicity need to analysed before the release of a GM crop. Proper policies and regulation on GM seed production and cultivation will increase the socio-economic conditions of farmers. The people should be educated about the GM crops and their safe use. Once adopted, the effect of GM crops on ecosystem need be monitored continuously, to prevent the emergence of new risk and contamination of native biodiversity. The use of GM crops so far has shown that unless a toxic gene is engineered, GM crops are completely safe to use. With proper regulation, surveillance and utilisation, this technique is a boom to the growing population.

## CONSENT FOR PUBLICATION

Not applicable.

## CONFLICT OF INTEREST

The authors confirm that this chapter contents have no conflict of interest.

## ACKNOWLEDGEMENTS

Declared none.

## REFERENCES

[1]     https://www.un.org/development/desa/en/news/population/world-population-prospects-2017.html

[2]     http://www.landcommodities.com/farmland-supply-and-investment-fundamentals/

[3]     Milford MH, Runge ECA. Beachell and Borlaug, two giants of the American Society of Agronomy's first century. Agron J 2007; 99: 595-8.
[http://dx.doi.org/10.2134/agronj2007.0004]

[4]     Oritz R, Mowbray D, Dowswell C, Rajaram S, Dedication-Norman E. Borlaug: The humanitarian plant scientist who changed the world. Plant Breed Rev 2007; 28: 1-37.

[5]     Rajaram S. Norman Borlaug: the man I worked with and knew. Annu Rev Phytopathol 2011; 49: 17-30.
[http://dx.doi.org/10.1146/annurev-phyto-072910-095308] [PMID: 21370972]

[6]     http://www.isaaa.org/resources/publications/briefs/53/

[7]     Jordan CF. Genetic Engineering, the Farm Crisis and World Hunger. BioSci 2002; 52(6): 523-9.
[http://dx.doi.org/10.1641/0006-3568(2002)052[0523:GETFCA]2.0.CO;2]

[8]     Clive J. Global status of commercialized biotech/GM Crops: 2013 ISAAA Brief No 46. Ithaca, NY: ISAAA 2013.

[9]     Jacobsen SE, Sorensen M, Pedersen SM, Weiner J. Feeding the world: genetically modified crops *versus* agricultural biodiversity. Agron Sustain Dev 2013; 33(4): 651-62.
[http://dx.doi.org/10.1007/s13593-013-0138-9]

[10]    Bawa AS, Anilakumar KR. Genetically modified foods: safety, risks and public concerns-a review. J Food Sci Technol 2013; 50(6): 1035-46.
[http://dx.doi.org/10.1007/s13197-012-0899-1] [PMID: 24426015]

[11]    Nordlee JA, Taylor SL, Townsend JA, Thomas LA, Bush RK. Identification of a Brazil-nut allergen in transgenic soybeans. N Engl J Med 1996; 334(11): 688-92.
[http://dx.doi.org/10.1056/NEJM199603143341103] [PMID: 8594427]

[12]    Butler D, Reichhardt T. Long-term effect of GM crops serves up food for thought. Nature 1999; 398(6729): 651-6.
[http://dx.doi.org/10.1038/19381] [PMID: 10227281]

[13]    Key S, Ma JKC, Drake PMW. Genetically modified plants and human health. J R Soc Med 2008; 101(6): 290-8.
[http://dx.doi.org/10.1258/jrsm.2008.070372] [PMID: 18515776]

[14]    Mercer DK, Scott KP, Bruce-Johnson WA, Glover LA, Flint HJ. Fate of free DNA and transformation of the oral bacterium *Streptococcus gordoni*i DL1 by plasmid DNA in human saliva. Appl Environ Microbiol 1999; 65(1): 6-10.

[http://dx.doi.org/10.1128/AEM.65.1.6-10.1999] [PMID: 9872752]

[15]   Knights D, Kuczynski J, Charlson ES, *et al.* Bayesian community-wide culture-independent microbial source tracking. Nat Methods 2011; 8(9): 761-3.
[http://dx.doi.org/10.1038/nmeth.1650] [PMID: 21765408]

[16]   Dubinsky EA, Butkus SR, Andersen GL. Microbial source tracking in impaired watersheds using PhyloChip and machine-learning classification. Water Res 2016; 105: 56-64.
[http://dx.doi.org/10.1016/j.watres.2016.08.035] [PMID: 27598696]

[17]   Smith A, Sterba-Boatwright B, Mott J. Novel application of a statistical technique, Random Forests, in a bacterial source tracking study. Water Res 2010; 44(14): 4067-76.
[http://dx.doi.org/10.1016/j.watres.2010.05.019] [PMID: 20566209]

[18]   Li LG, Yin X, Zhang T. Tracking antibiotic resistance gene pollution from different sources using machine-learning classification. Microbiome 2018; 6(1): 93.
[http://dx.doi.org/10.1186/s40168-018-0480-x] [PMID: 29793542]

[19]   Healy S. notes 2006; 35: 211-2.

[20]   Kendall Bioengineering of crops (Report of the World Bank Panel on Transgenic Crops). Washington, DC, USA: World Bank 1997.

[21]   Gupta K, Karihaloo JL, Khetarpal RK. Biosafety regulations of Asia-Pacific Countries Asia-Pacific Association of Agricultural Research Institutions, Bangkok. Rome, UK: Asia-Pacific Consortium on Agricultural Biotechnology, New Delhi and Food and Agricultural Organization of the United Nations 2008.

[22]   Mascia PN, Flavell RB. Safe and acceptable strategies for producing foreign molecules in plants. Curr Opin Plant Biol 2004; 7(2): 189-95.
[http://dx.doi.org/10.1016/j.pbi.2004.01.014] [PMID: 15003220]

[23]   Weil JH. Are genetically modified plants useful and safe? IUBMB Life 2005; 57(4-5): 311-4.
[http://dx.doi.org/10.1080/15216540500092252] [PMID: 16036615]

[24]   Daniell H. GM crops: public perception and scientific solutions. Trends Plant Sci 1999; 4(12): 467-9.
[http://dx.doi.org/10.1016/S1360-1385(99)01503-4] [PMID: 10562729]

[25]   Cban Are GMOs better for the environment, Report 2, Canadian Biotechnology Active Network 2015.

[26]   Scheffler JA, Parkinson R, Dale PJ. Evaluating the effectiveness of isolation distances for field plots of oilseed rape (*Brassica napus*) using a herbicide-resistance transgene as a selectable marker. Plant Breed 1995; 114: 317-21.
[http://dx.doi.org/10.1111/j.1439-0523.1995.tb01241.x]

[27]   Metz M, Fütterer J. Biodiversity (Communications arising): suspect evidence of transgenic contamination. Nature 2002; 416(6881): 600-1.
[http://dx.doi.org/10.1038/nature738] [PMID: 11935144]

[28]   Ortiz-García S, Ezcurra E, Schoel B, Acevedo F, Soberón J, Snow AA. Absence of detectable transgenes in local landraces of maize in Oaxaca, Mexico (2003-2004). Proc Natl Acad Sci USA 2005; 102(35): 12338-43.
[http://dx.doi.org/10.1073/pnas.0503356102] [PMID: 16093316]

[29]   Reichman JR, Watrud LS, Lee EH, *et al.* Establishment of transgenic herbicide-resistant creeping bentgrass (*Agrostis stolonifera* L.) in nonagronomic habitats. Mol Ecol 2006; 15(13): 4243-55.
[http://dx.doi.org/10.1111/j.1365-294X.2006.03072.x] [PMID: 17054516]

[30]   Oliver MJ. Why we need GMO crops in agriculture, science of medicine. Natl Rev 2014; 111(6): 493-507.

[31]   Losey JE, Rayor LS, Carter ME. Transgenic pollen harms monarch larvae. Nature 1999; 399(6733): 214.
[http://dx.doi.org/10.1038/20338] [PMID: 10353241]

[32]   Hails RS. Genetically modified plants - the debate continues. Trends Ecol Evol (Amst) 2000; 15(1): 14-8.
[http://dx.doi.org/10.1016/S0169-5347(99)01751-6] [PMID: 10603498]

[33]   Fontes. The environmental effects of genetically modified crops resistant to insects. Neotrop Entomol 2002; 31: 497-513.
[http://dx.doi.org/10.1590/S1519-566X2002000400001]

[34]   Naranjo S. Impacts of Bt Crops on non-target invertebrates and insecticide use patterns. Perspect Agric Vet Sci Nutr Nat Resour 2009; 4: 1-11.
[http://dx.doi.org/10.1079/PAVSNNR20094011]

[35]   Singh M. Biosafety concerns and regulatory framework for transgenics. Research Journal of Agriculture and Forestry Sciences 2014; 2: 7-13.

[36]   Terefe M. Biosafety issues of genetically modified crops: Addressing the potential risks and the status of GMO crops in ethiopia. Clon Transgen 2018; 7: 164.
[http://dx.doi.org/10.4172/2168-9849.1000164]

[37]   Anderson K, Yao S. GMOs and world trade in agricultural and textile products. Pac Econ Rev 2003; 8: 157-69.
[http://dx.doi.org/10.1111/j.1468-0106.2002.00217.x]

[38]   Council for agricultural science and technology (CAST). The potential impacts of mandatory labelling for genetically engineered food in the United States Issue Paper 54 CAST. Ames, Iowa: Available 2014.

[39]   Hemphill TA, Banerjee S. Mandatory Food Labeling for GMOs. Regulation 2014; 2014–2015: 7-10.

[40]   Han Y, *et al.* Evolution of GMO Regulations in the European Union and Indications for China. Zhejiang Agric Sci 2013; 11: 1482-9.

[41]   Yang W. Regulation of genetically modified organisms in china. Rev Eur Community Int Environ Law 2003; 12(1): 99-108.
[http://dx.doi.org/10.1111/1467-9388.00347]

[42]   Hino A. Safety assessment and public concerns for genetically modified food products: The Japanese experience. Toxicol Pathol 2002; 30(1): 126-8.
[http://dx.doi.org/10.1080/01926230252824815] [PMID: 11890464]

[43]   Ebata A, Punt M, Wesseler J, *et al.* For the approval process of GMOs: the Japanese case. AgBioForum 2013; 16(2): 140-60.

[44]   Ministry of Environment, Forest and Climate Change (MoEFCC), GSR 1037 (E) Rules for manufacture, use/import/export & storage of hazardous microorganisms/genetically engineered organisms or cells 1989.

[45]   Ahuja V. Regulation of emerging gene technologies in India. BMC Proc 2018; 12 (Suppl. 8): 14.
[http://dx.doi.org/10.1186/s12919-018-0106-0] [PMID: 30079105]

**CHAPTER 7**

# Pesticide Pollution

**A. Anitha**[1,*] and **S. Geethalakshmi**[2]

[1] *Department of Biotechnology, Nehru Arts and Science College, Coimbatore 641 105,Tamil Nadu, India*

[2] *Department of Biotechnology, Sree Narayana Guru College, Coimbatore 641 105, Tamil Nadu, India*

**Abstract:** Pesticides are substance or mixture of substance which differ in their physical, chemical and identical properties from one to other. Hence, the pesticides are classified based on these properties. The classification of pesticides is based on (i) mode of entry, (ii) pesticide function and the pest organism they kill, and (iii) on the chemical composition. Pesticides are formulated in various forms like Liquids, Powders, Granules, Baits, Dust, Smoke generators, Ultra Low Volume liquids, *etc*. They are chemicals that are used to kill or control pests. Despite beneficial results of using pesticides in agriculture and public health sector, their use also invites deleterious environmental and public health effects. It has been observed that inappropriate application of pesticides may adversely affect every component of the environment. Due to the intensive application of pesticides, pests may evolve to develop resistance. Biological controls, such as resistant plant varieties and the tradition of pheromones, have been fruitful and, at periods, eternally resolve a pest problem. The most serious effects involve the destruction of non-target pest organisms (earthworm, pollinator and predators), loss in biological diversity, microbial diversity, and soil biomass or community assembly. These ecological losses owing to pesticides application are economically or socially significant. Hence, pesticides user, especially farmer, is suggested to reduce the impacts of pesticides by minimizing their application or by replacing it with bio-pesticides.

**Keywords:** Classification, Health and Environmental Concern, Pollution, Pesticides.

## INTRODUCTION

In the past three eras, there has been an increasing global concern over the community health influences accredited to environmental pollution. Industrial revolution that leads to environmental pollution and populations of developing countries  are particularly vulnerable to  toxic substances resulting from industrial

---

[*] **Corresponding author Dr. A. Anitha:** Department of Biotechnology, Nehru Arts and Science Collège, Coimbatore - 641 105, Tamil Nadu, India; Tel: +91 99943 15759; E-mail: anithavarshini22@gmail.com

**J. Senthil Kumar, P. Ponmurugan & A. Vinothkanna (Eds.)**

progressions. Pollution is defined as the introduction of contaminants into the natural environment that causes adversative fluctuations or discomfort in living or non-living belongings or might impair the environment. Pollutants in the machineries of pollution, can be either foreign substances such as chemicals, toxins, drugs, or energies, like heat, light, or noise or naturally occurring contaminants, *i.e.*, natural constituents from its environment for better or worse.

In terms of eco-system, classification of the pollutants has been divided into two basic groups: Biodegradable Pollutants & Non-Biodegradable Pollutants.

## BIODEGRADABLE POLLUTANTS

Biodegradable pollutants can be broken down and handled by living organisms, counting as organic waste products, phosphates, and inorganic salts. Non-biodegradable pollutants may not be decomposed by living organisms and consequently persist in the ecosphere intended for extremely extended periods of time. They contain metals, plastics, glass, pesticides, and radioactive isotopes [1]. In 2015, pollution killed 9 million people in the world [2].

Depending upon the nature and its interaction with the environment, the pollution caused by diverse pollutants can be classified into Air pollution, light pollution, littering, noise pollution, plastic pollution, soil contamination, radioactive contamination, thermal pollution, visual pollution, and water pollution. In recent years, people have been exposed to numerous classes of constituents with a wide-range of spectrum owing to the hastily evolving technology. Technology has brought us strong services, and thousands of chemicals fashioned in different extents which remain up on the market each year. One of these chemical substances are pesticides [3].

### What are Pesticides

Pesticides are a natural (Bacteria, Viruses), synthetic (Organic or Inorganic) or mixture of substances intended for preventing, destroying, repelling, or lessening the damage of living creatures caused by any pest. Although the term pesticide has been frequently associated with synthetic chemical compounds, it was not until comparatively that synthetic pesticides initiated into use [4]. Synthetic pesticides are chemicals, made by humans.

### *History*

Pesticides are purposely applied to the environment with the purpose to overwhelm pests and to protect agricultural products. Researches during the late 19th and initial 20th periods permitted the human beings to develop contemporary pesticides. The unwanted organism remained controlled using novel mixtures with accurate proportion.

By the late 19th century, U.S. farmers used calcium arsenate, nicotine sulfate, and sulfur to control insect pests in field crops and till the middle of 20th century. Ancient Romans controlled the weeds by salt and eradicated the insect pests by burning sulfur recognized as brimstone. In the 1600s, ants were controlled through the mixtures of honey besides arsenic. Early plant-derived insecticides included nicotine to control aphids, hellebore to control body lice, and pyrethrins to control a wide variety of insects [5].

The new era of pest management was started in 1945 for civilian/agricultural usage, subsequently the accessibility of dichlorodiphenyltrichloroethane (DDT). DDT was particularly favored for its broad-spectrum activity in contradiction of agricultural insect pests [6]. Unfortunately, persistence of DDT ended a deprived choice for practice in agriculture after World War II.

Except DDT, new-fangled chemicals like aldrin, BHC, endrin, dieldrin, and 2,4-D came into practice after World War II due to the aforementioned effective and inexpensive nature. Continuous tradition of pesticides created pest's resistance correspondingly damaged non-target plants in addition to animals. Rachel Carson's book, Silent Spring in 1962, shook public confidence in pesticide usage [7].

Many pesticides are not easily degradable, they persist in soil, leach to groundwater and surface water and contaminate wide environment. Depending on their chemical properties, they can go in the organism, bioaccumulate in food chains and subsequently influence human health [8]. Globally, intensive pesticide application results in numerous negative effects in the atmosphere, injury of biodiversity and global ecological degradation.

## Toxicological Classification of Pesticides

Pesticides is an umbrella term that includes several classes of insecticides, herbicides, fungicides, rodenticides, wood preservatives, garden chemicals and household disinfectants used to either kill or protect the plants and animals from pests. These pesticides differ from each other by their physical, chemical and undistinguishable properties from one class to other.

Therefore, the classification of pesticides has been based on their properties depending on the needs. Currently, the classification of pesticides has been suggested by Drum [9]. The three most prevalent methods of pesticides modules comprise: (i) classification based on the mode of entry, (ii) classification based on pesticide function and the pest organism they kill, and (iii) classification based on the chemical composition of the pesticide [9].

## *Classification Based on Mode of Entry*

It defines the way pesticides come and interact with or enter the target. The entry may include systemic, contact, they work *via* the skin; stomach poisons, they have to be eaten; fumigants, they produce a vapor that kills organisms; and repellents, is a substance applied to skin, clothing, or other surfaces which discourages insects from landing or climbing on that surface [9].

## *Classification Based on Pesticide Function and Pest Organism they Kill*

Pesticides are classified based on the target pest's organism and are named based on their activity. The name arises from the Latin word *'cide'* means kill or killer, used as a suffix of pests which they kill/encountered. For instance, insecticides stand pesticides that target insects, in addition, herbicides stand plants targeted. The others are rodenticides, a poison used to kill rodents; fungicides, a chemical that destroys fungus; acaricides and miticides, a substance poisonous to mites or ticks; molluscicides, are pesticides contrary to molluscs; bactericides, a material which slays bacteria; avicides, which can be cast-off to kill birds and virucides, that neutralizes or abolishes viruses.

Other class of pesticides are according to their function. For examples: growth regulators, which stimulate or retard the pests growth; defoliants, cause abscission of plants; desiccants, drying of plants by a hygroscopic for the mechanical harvest of plants or cause insects to dry out and die; repellents which repel pests; attractants, that attract pests, usually to trap the pest; and chemo sterilant, is a chemical compound that responses in the reproductive sterility.

## *Classification Based on Chemical Composition of Pesticides*

The most communal and valuable method of classifying pesticides is grounded on their chemical configuration and nature of vigorous ingredients. Such classification gives a clear idea of the effectiveness, physical and chemical possessions of the respective pesticides. The information on chemical and physical characteristics of pesticides is very beneficial in defining the means of application, precautions that are necessisary to be engaged during application and its rates. Based on the chemical configuration, pesticides are classified into four foremost groups *viz.*; organochlorines, organophosphorus, carbamates and pyrethrin and pyrethroids (Table **1**) [4].

## Organochlorines

Chemicals like organochlorine belonged to organic compounds with more than five chlorine atoms. They are the first synthetic organic pesticides, which are used in public health and in agriculture. Organo chlorine acts as disruptors of the nervous system leading to convulsions and paralysis of the insect and its eventual death. They can cause serious endocrine disorders in mammals, fish and birds, therefore most of the organo chlorine family have been banned worldwide, in agriculture [10].

## Organophosphates

Organophosphates are another type of highly toxic pesticides that contains phosphate group and occupied 48.6% of all existing pesticides in 1997 [11]. These chemical compounds constrain the acetyl cholinesterase enzyme, which hydrolyses acetylcholine in the nervous system of abundant species, together with humans [12]. Organophosphates are easier to degrade than organochlorines, but residues are the biggest threats to the eco-system and food industry because of its irreversible acute toxicity. Every twelve months, acute poisoning has been reported amongst 3 million cases of pesticide acquaintance, resulting in the deaths of human populations each year [13].

## Carbamates

Carbamates are organic pesticides and the derivative of carbamic acid. These are reversibly deactivating the carbamates dependent acetylcholinesterase enzyme [14].

## Pyrethrin and Pyrethroids

Pyrethroids are synthetic equivalents of the naturally occurring pyrethrins, a product of pyrethrum plant flowers (*Chrysanthemum cinerariaefolium*), tomimic the insecticidal action of the natural pyrethrum. They are heterocyclic complexes usually less toxic while associated with three generations of outdated highly toxic organochlorines, organophosphates and carbamates.

Pyrethroids are acknowledged for their fast knocking down effect against insect pests, facile biodegradation, and low mammalian toxicity; nevertheless, they are extremely toxic to aquatic creatures, for instance, mollusks, fish, and arthropods [15].

**Table 1. Class of Pesticides.**

| Class of Pesticides | Examples | Area of Effect |
|---|---|---|
| **Organochlorines** | DDT, lindane, endosulfan, aldrin, dieldrin and chlordane | Reproductive, endocrine, nerve and immune system |
| **Organophosphorus** | parathion, malathion, diaznon and glyphosate | Central nerves system |
| **Carbamates** | carbaryl, carbofuran, propoxur and aminocarb | Central nerves system |
| **Pyrethrin and pyrethroids** | Cypermethrin and Permethrin | Poorly understood |

Except for these classifications, pesticides are classified according to the mode of formulation, activity spectrum, and toxicity level.

How is it formulated?

Pesticides are formulated in various forms like Liquids, Powders, Granules, Baits, Dust, Smoke generators, Ultra Low Volume (ulv) liquids, *etc.*

About the target range?

Comprehensive spectrum pesticides (chemicals that kill a wide range of pests).

Discriminating pesticides (chemicals that kill only a specific pest or group of pests).

To what toxicity class does it belong?

Another way of grouping pesticide is in accordance with potential hazards to human health. The World Health Organization (WHO) has developed the classes for chemical pesticides according to their toxicity:

Class Ia = Extreme hazardous in nature.
Class Ib = High hazardous in nature.
Class II = Moderate hazardous in nature.
Class III = Slight hazardous in nature.
Class IV = Product unlikely to present acute hazard in normal use [16].

## Pesticide Pollution

Pesticides are defined as the chemicals that are pertained to kill otherwise control pests. This includes herbicides (for getting rid of weeds), insecticides (for treating fungicides), nematocides (for controlling nematodes) as well as rodenticides (for treating vertebrate poisoning). Despite beneficial results by means of pesticides in agriculture and public health segment, their use also calls deleterious environmental and public health impacts. Environmental pollution mediated by the pesticides occurs as soon as the heavy wind or rain falls on the aforementioned lands, spreading the pesticides, being toxic chemicals, into unintended areas, coming in contact with natural resources such clean air, water, land, plants, and animals, thereby contaminating or harming them.

## How We are Exposed to Pesticides

Mode of entry of pesticides in the human body is through (i) inhalation of polluted air, dust and vapor that contain pesticides; (ii) oral exposure by consuming contaminated food and water; and (iii) dermal exposure by direct contact with pesticides [17].

## How Pesticides Contaminate Groundwater

Pesticide contamination of groundwater is an issue of national importance for the reason that groundwater is intended for drinking by about 50 out of a hundred of the Nation's population. This especially worries about the people living in the agricultural zones where pesticides are recorded most often, as about 95 out of a hundred of the population depend on groundwater for drinking. Pesticides are sprayed onto food, especially fruits and vegetables, they secrete into soils and groundwater, which can end up in drinking water and pesticide spray can drift and pollute the air. In addition, the pesticides can spread to the water-bearing aquifers below ground from applications onto crop fields, seepage of contaminated surface water, accidental spills and leaks, improper disposal, and even through injection of waste material into wells (Fig. 1).

## POTENTIAL IMPACT ON HUMAN HEALTH

Toxicity of pesticides depends on the nature of toxicants, routes of exposure (oral, dermal and inhalation), dose and organism. Toxicity can be either acute or chronic. Acute toxicity is the ability of a substance to cause harmful effects that develop rapidly following absorption, *i.e.*, a few hours or a day. Chronic toxicity is the capability of a substance to cause adverse well-being effects resulting from long-term exposure to a constituent.

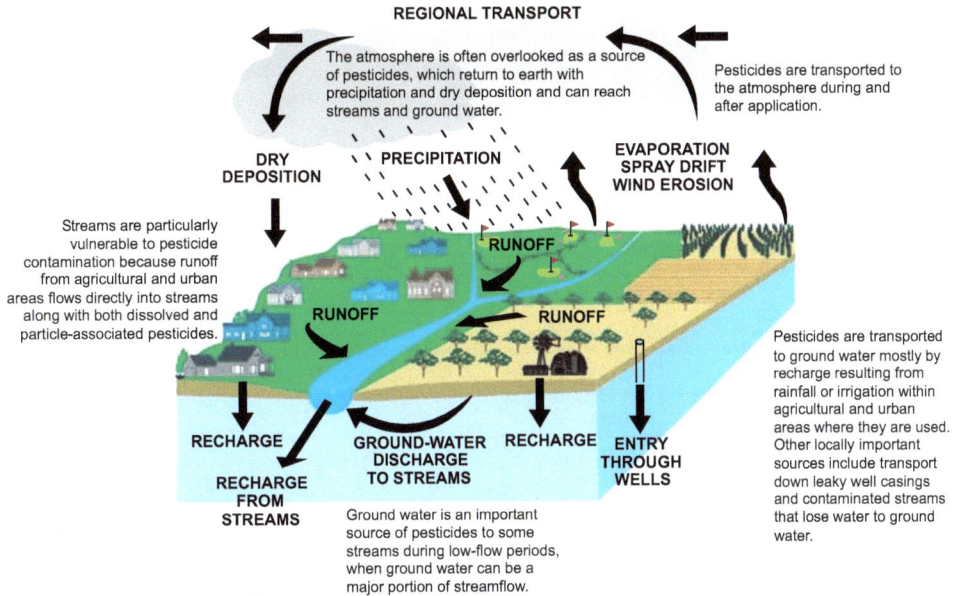

**REGIONAL TRANSPORT**

The atmosphere is often overlooked as a source of pesticides, which return to earth with precipitation and dry deposition and can reach streams and ground water.

Pesticides are transported to the atmosphere during and after application.

**DRY DEPOSITION**    **PRECIPITATION**

**EVAPORATION SPRAY DRIFT WIND EROSION**

Streams are particularly vulnerable to pesticide contamination because runoff from agricultural and urban areas flows directly into streams along with both dissolved and particle-associated pesticides.

**RUNOFF**

**RUNOFF**    **RUNOFF**

Pesticides are transported to ground water mostly by recharge resulting from rainfall or irrigation within agricultural and urban areas where they are used. Other locally important sources include transport down leaky well casings and contaminated streams that lose water to ground water.

**RECHARGE**

**GROUND-WATER DISCHARGE TO STREAMS**   **RECHARGE**   **ENTRY THROUGH WELLS**

**RECHARGE FROM STREAMS**

Ground water is an important source of pesticides to some streams during low-flow periods, when ground water can be a major portion of streamflow.

**Fig. (1).** Pesticide transportation flow chart.

## Acute Effect

Single exposure may produce a harmful effect by numerous routes of entry that are labelled "acute effects". The dissimilar routes of exposure are dermal (skin), inhalation (lungs), oral (mouth), and the eyes. More than a few symptoms of acute illness are body aches, headaches, skin rashes, nausea, poor concentration, dizziness, cramps, impaired vision, panic attacks, and in severe cases, coma and death, which could occur due to pesticide poisoning.

## Chronic Effect

Small doses repeated over a period of time lead to harmful effects that are termed "chronic effects." Chronic effects from exposure to certain pesticides include birth defects, toxicity to a fetus, and production of benign or malignant tumors, genetic changes, blood disorders, nerve disorders, endocrine disruption, and reproduction effects. Continuous and repeated exposure to sub-lethal quantities of pesticides for a long period of time (may be several years to decades), causes chronic illness in humans [18]. More commonly, agricultural farmer is at a higher risk to be affected and symptoms may appear at a later stage. Recently, several studies establish a connection between pesticide exposure and the incidences of human chronic diseases distressing reproductive, renal, nervous, cardiovascular, and respiratory systems [19]. Some of the most common chronic diseases due to long exposure to pesticides are given in Table **2**.

**Table 2. Public chronic diseases based on pesticides [17].**

| Diseases | References |
|---|---|
| Cancer (Childhood and adult brain cancer Renal cell cancer; Lymphocytic Leukemia(CLL); Prostate Cancer) | Band *et al.* [20]; Cocco*et al.* [21], |
| Parkinson Alzheimer disease (Neuro degenerative diseases) | Hayden *et al.* [22]; Tanner *etal.*, [23] |
| Artery disease (Cardio-vascular disease) | Abdullah *et al.* [24]; Andersen *et al.* [25], |
| Type 2 Diabetes | Son *et al.* [26], |
| Reproductive disorders | Greenlee *et al.* [27], |
| Birth defects | Winchester *et al.* [28]; Mesnage*et al.* [29], |
| Infertility and breast pain (Hormonal imbalances) | Xavier *et al.* [30], |
| Respiratory diseases | Chakraborty *et al.* [31]; Hoppin*et al.* [32], |

## IMPACTS ON ENVIRONMENT

Extensive application and subsequent disposal of pesticides by farmers and the general public lead to pesticide accumulation/pollution in the environment. Almost all the areas will be affected by pesticides. Pesticides released into the atmosphere have different fates. When pesticides are sprayed to the agricultural crop, it finds their way for its spread in the air, is absorbed in the soil, or dissolves in the water and eventually ends up in further segments of the environment. Besides, pesticides that are applied straight to the soil might be eroded off and spreads to nearby superficial water bodies through shallow runoff or may percolate over the soil to inferior soil layers and groundwater. The effects of pesticides on the environmental system may range from minor deviation on the normal functioning of the eco-system to the loss of species diversity. For example, most organochlorine pesticides are persistent in the environment for a long time, hence, resulting in eggshell thinning in raptorial birds, thyroid disturbance in rodents, birds, amphibians and fish (Table **3**).

**Table 3. Pesticide Environmental effects.**

| Pesticide/Class | Effect(s) |
|---|---|
| DDT/Diclofol, Dieldrin | Juvenile population decline and adult mortality in wildlife reptiles. |
| Organochlorine DDT/DDE | Egg shell thinning in raptorial birds, Thyroid disruption properties in rodents, birds, amphibians and fish. |
| DDT | Carcinogen [33]. |
| Toxaphene | Juvenile population decline and adult mortality in wildlife reptiles. |

*(Table 3) cont.....*

| Pesticide/Class | Effect(s) |
|---|---|
| Triazine | Earthworms became infected with monocystid gregarines [34]. |
| Chlordane, Carbamates, the phenoxyherbicide 2,4-D, and atrazine | Interact with vertebrate immune systems [35]; Thyroid disruption properties in rodents, birds, amphibians and fish [36]. |
| Anticholinesterase | Bird poisoning [37]and Animal infections [38]. |
| Pyrethroid, Thiocarbamate, Triazine | Thyroid disruption properties in rodents, birds, amphibians and fish [36]. |
| Neonicotinoic/Nicotinoid | respiratory, cardiovascular, neurological, and immunological toxicity in rats and humans [39]. |
| Imidacloprid, Imidacloprid/pyrethroid λ-cyhalothrin | Impaired foraging, brood development, and colony success in terms of growth rate and new queen production [40]. |
| Thiamethoxam | High honey bee worker mortality due to homing failure [41] [42]. |

## Impact on Plants

Nitrogen fixation is inherited in soil by pesticide, which is mandatory for the development of higher plants. The insecticides DDT, methyl parathion, and especially pentachlorophenol have been revealed to restrict the legume-rhizobium chemical signalling [43]. The lessening of this symbiotic chemical messenger leads to nitrogen fixation reduction and consequently diminishes crop yields.

Pesticides can implement the bees and are powerfully implicated in pollinator worsening, the damage of species that pollinate plants, counting through the machinery of Colony Collapse Disorder [44, 45]. The application of pesticides to blooming crops can destroy honeybees [46], perform as pollinators. On the other side, pesticides have about straight injurious impact on the plant, including deprived root hair expansion, shoot yellowing and shortened plant growth [47].

## Impact on Soil

The widespread use of pesticides in agricultural exercise can destroy and damage the microbial community breathing in the soil, predominantly through the time of overused. The consequences of pesticides on soil microorganisms are obstructed by the perseverance, attentiveness, and poisonousness of the applied pesticide, in addition to innumerable environmental factors. This multipart collaboration of factors brands is problematic to draw significant conclusions around the communication of pesticides with the soil eco-system. Over-all, long-term pesticide applications can interrupt nutrient cycling [48].

The general biodiversity in the soil is decrease by the use of pesticides. Not using the chemicals results in higher soil quality, with the additional effect that more organic matter in the soil allows for higher water retention. This helps up regulate the yields for farms in drought centuries, when organic farms had yields 20-40 out of hundred higher than their conventional corresponding item. A slighter content of organic matter in the soil increases the quantity of pesticide that will leave the zone of submission, since organic matter binds to and helps break down pesticides [49].

## Impact on Aquatic Life

Pesticide-contaminated water may affect the fish and other aquatic biotas [50]. Pesticides on the superficial runoff into rivers and watercourses can be fatal to aquatic life [51]. Application of herbicides to aquatic bodies can cause the death of fish when the water's oxygen utilization by the dead plants utilize for decay, in turn, causes fish suffocation. Herbicides like copper sulfite that are applied to kill water plants are toxic to all the available aquatic animals. Frequent exposure to sublethal doses of particular pesticides can cause physiological and behavioral deviations in fish that diminish the fish population, such as abandonment of shells and broods, decreased immunity against the disease, and predator avoidance also decreased [50].

In addition, pesticides accrued in water bodies destroy off zooplankton, which is the chief source of nutrition for undeveloped fish [49]. Pesticides can also kill off insects on which around fish feed, triggering the fish to transport beyond in exploration of food and revealing them to superior risk as of predators [50].

## Impacts on Water and Air Ecosystem

Pesticide residues in water are a foremost concern as they stood a serious hazard to biological groups, which includes humans, animals' plants, *etc*. Pesticides enter into the water by means of accidental spillage, effluent from industry, surface runoff and carriage from pesticide-treated soils, spray equipment washing after spray operation, drift into ponds, lakes, streams and river water, aerial spray to control water inhibiting pests [52].

Equally, the occurrence of pesticides in the air can be caused by numerous factors, including spray drift, volatilization from the treated surfaces, and aerial application of pesticides. Degree of drift subjected to droplet size and wind speed. The rate of volatilization is dependent on time after pesticide treatment, the surface on which the pesticide settles, the ambient temperature, humidity and wind speed and the vapor pressure of the ingredients. The nature of the pesticide compounds, *i.e.*, volatility or semi-volatility correspondingly constitutes a significant hazard of atmospheric pollution of huge cities [53].

## Impacts on Soil Micro-Flora

A major portion of the pesticides (non-target) applied in agriculture and other sources possibly will accumulate in the soil. Further, the indiscriminate use of pesticides worsens this soil accumulation. Soil properties and microflora get affected due to pesticides, which may undergo a process like degradation, transportation, and adsorption/desorption [54]. The degraded pesticides interact with the soil and its indigenous microorganisms, therefore altering its microbial diversity, biochemical reactions and enzymatic activity [54, 55]. Soil eco-system and loss of soil fertility are disturbed because of alteration in the microbial diversity and soil biomass. Pesticide applications may also inhibit or kill a certain group of microorganisms and outnumber other groups by freeing them as of the competition [54]. Furthermore, pesticides adversely affect the soil's dynamic biochemical reactions, counting nitrogen fixation, nitrification, and ammonification by activating/deactivating specific soil microorganisms and/or enzymes [54, 55]. Pesticides influence soil organic matter mineralization, which determines the soil quality and productivity.

## Impacts on Non-Target Organism

Most of the applied pest killers adversely affect non-target organisms such as earthworm, natural predators and pollinator. Unfortunately, natural predators such as parasitoids and predators (essential for controlling pest population level) are most susceptible to insecticides and are harshly affected [56] that exacerbate pest problems. Due to the absence of natural enemies, added insecticide sprays stands are required to control the target pest.

Pollinators such as bees, fruit flies, some beetles, and birds can be used as bioindicators of the eco-system, and their activities are affected by environmental stress caused by pesticide application and habitat modifications [57]. Insect pollinator's loss is directly affected by the use of pesticides, and corps are affected indirectly because of the inadequate pollinators [58].

## Pest Resistance

Pests may evolve to develop resistance to pesticides. Initially, all the pests show susceptibility to pesticides; continuous application leads to mutations in their genetic makeup that ended with resistance in pest and survived to reproduce. Resistance can be generally achieved over pesticide alternation, which involves alternating among pesticide classes with dissimilar modes of exploit to postpone the onset of or alleviate the existing pest resistance.

## Eradicating Pesticides

Countless alternatives exist to diminish the effects of pesticides present in the environment. Alternatives which comprise intensive removal through labour, placing traps and lures, heat smearing, weeds covering with plastic, removal of pest breeding sites, soil health maintenance that breed healthy, added resistant plants and native species cropping, are certainly more resistant to inherent pests and supporting birds and other pest predators as biocontrol agents.

Resistant plant varieties and pheromones usage have been successful in relation to biological control and at eras everlastingly resolve a pest problem. Integrated Pest Management (IPM) relies upon chemicals when other alternatives are ineffective; also, it causes less harm to humans and the environment. Biotechnology tools play an innovative role in pests control. Genetically modified (GM) strains can be used to intensify their pest resistance correspondingly to increase pesticide resistance.

## SUMMARY

Pesticides have been classified based on numerous criteria. The most prevalent source of pesticide cataloguing is based on the mode of entry, pesticide function and the pest organism they kill and chemical composition. Pesticides with parallel structures have similar characteristics and usually share a common mode of action. The active ingredients of the pesticides are either inorganic or organic pesticides. It has been observed that the inappropriate application of pesticides may adversely affect all levels of biological organization and every component of the environment. The effects can be universal or resident, temporary or permanent or transitory (acute) or long-standing(chronic). The most serious effects involve the destruction of non-target pest organisms (earthworm, pollinator and predators), loss in biological diversity, microbial diversity, and soil biomass or community structure. These ecological losses due to pesticides application are economically or socially important. Hence, pesticides user, especially farmer, is suggested to reduce the impacts of pesticides by minimizing their application or by replacing it with bio-pesticides.

## CONSENT FOR PUBLICATION

Not applicable.

## CONFLICT OF INTEREST

The authors confirm that this chapter contents have no conflict of interest.

## ACKNOWLEDGEMENTS

Declared none.

## REFERENCES

[1]     Santos MA. Managing Planet Earth: Perspectives on Population, Ecology, and the Law. Westport, Connecticut: Bergin & Garvey 1990; p. 44.

[2]     Laura Beil. Pollution killed 9 million people in 2015. Sciencenews.org.

[3]     Erdogmuş SF, Eren Y, Akyıl D, Ozkara A, Konuk M, Saglam E. Evaluation of *in vitro* genotoxic effects of benfuracarb in human peripheral blood lymphocytes. Fresenius Environ Bull 2015; 24(3): 796-9.

[4]     Eldridge BF. Pesticide application and safety training for applicators of public health pesticides. Vector-Borne Disease Section, Sacramento, CA,: California Department of Public Health 2008.

[5]     Fishel FM. Pest Management and Pesticides: A Historical Perspective 2013.

[6]     Felsot AS. Pesticides & Health—Myths vs Realities. New York, NY: American Council on Science and Health 2006; p. 107.

[7]     Gribble GW. Naturally occurring organohalogen compounds. Acc Chem Res 1998; 31: 141-52.
[http://dx.doi.org/10.1021/ar9701777]

[8]     Kogan M. Integrated pest management: historical perspectives and contemporary developments. Annu Rev Entomol 1998; 43: 243-70.
[http://dx.doi.org/10.1146/annurev.ento.43.1.243] [PMID: 9444752]

[9]     Drum C. Soil Chemistry of Pesticides. USA: PPG Industries, Inc. 1980.

[10]    Willet KL, Ulrich EM, Hites RA. Differential toxicity and environmental fates of hexachlorocyclohexane isomers. Environ Sci Technol 1998; 32: 2197-207.
[http://dx.doi.org/10.1021/es9708530]

[11]    Zhang Y. New Progress in Pesticides in the World. Beijing: Chemical Industry Press 2007.

[12]    Zahran MM, Abdel-Aziz KB, Abdel-Raof A, Nahas EM. The effect of subacute doses of organophosphorus pesticide, Nuvacron, on the biochemical and cytogenetic parameters of mice and their embryos. Res J Agric Biol Sci 2005; 1: 277-83.

[13]    Marrs TC. Organophosphate poisoning. Pharmacol Ther 1993; 58(1): 51-66.
[http://dx.doi.org/10.1016/0163-7258(93)90066-M] [PMID: 8415873]

[14]    Morais S, Correia M, Domingues V, Delerue-Matos C. Urea pesticides. In: Stoytcheva M, Eds. Pesticides-Strategies for Pesticides Analysis. London, SW7 2QJ, UK: IntechOpen Limited 2011; pp. 241-62.
[http://dx.doi.org/10.5772/13126]

[15]    Zheng S, Chen B, Qiu X, Chen M, Ma Z, Yu X. Distribution and risk assessment of 82 pesticides in Jiulong River and estuary in South China. Chemosphere 2016; 144: 1177-92.
[http://dx.doi.org/10.1016/j.chemosphere.2015.09.050] [PMID: 26461443]

[16]    Tano ZJ. Identity, physical and chemical properties of pesticides. In: Stoytcheva M, Eds. Pesticides in the Modern World - Trends in Pesticides Analysis. London, SW7 2QJ, UK: IntechOpen Limited 2011; pp. 1-18.

[17]    Yadav IC and Devi NL. Pesticides classification and its impact on human and environment. Environ Sci &Engg 2017; 6: 140-58.

[18]    Shim YK, Mlynarek SP, van Wijngaarden E. Parental exposure to pesticides and childhood brain cancer: U.S. Atlantic coast childhood brain cancer study. Environ Health Perspect 2009; 117(6): 1002-

6.
[http://dx.doi.org/10.1289/ehp.0800209] [PMID: 19590697]

[19]    PAN. Pesticides and health hazards Facts and figures, Pesticide Action Network. Germany: GLS Gemeinschaftsbank 2012.

[20]    Band PR, Abanto Z, Bert J, *et al*. Prostate cancer risk and exposure to pesticides in British Columbia farmers. Prostate 2011; 71(2): 168-83.
[http://dx.doi.org/10.1002/pros.21232] [PMID: 20799287]

[21]    Cocco P, Satta G, Dubois S, *et al*. Lymphoma risk and occupational exposure to pesticides: results of the Epilymph study. Occup Environ Med 2013; 70(2): 91-8.
[http://dx.doi.org/10.1136/oemed-2012-100845] [PMID: 23117219]

[22]    Hayden KM, Norton MC, Darcey D, *et al*. Occupational exposure to pesticides increases the risk of incident AD: the Cache County study. Neurology 2010; 74(19): 1524-30.
[http://dx.doi.org/10.1212/WNL.0b013e3181dd4423] [PMID: 20458069]

[23]    Tanner CM, Kamel F, Ross GW, *et al*. Rotenone, paraquat, and Parkinson's disease. Environ Health Perspect 2011; 119(6): 866-72.
[http://dx.doi.org/10.1289/ehp.1002839] [PMID: 21269927]

[24]    Abdullah NZ, Ishaka A, Samsuddin N, Mohd RR, Mohamed AH. Chronic organophosphate pesticide exposure and coronary artery disease: Finding a bridge, IIUM Research, Invention and Innovation Exhibition. IRIIE 2011; p. 223.

[25]    Andersen HR, Wohlfahrt-Veje C, Dalgård C, *et al*. Paraoxonase 1 polymorphism and prenatal pesticide exposure associated with adverse cardiovascular risk profiles at school age. PLoS One 2012; 7(5): e36830.
[http://dx.doi.org/10.1371/journal.pone.0036830] [PMID: 22615820]

[26]    Son HK, Kim SA, Kang JH, *et al*. Strong associations between low-dose organochlorine pesticides and type 2 diabetes in Korea. Environ Int 2010; 36(5): 410-4.
[http://dx.doi.org/10.1016/j.envint.2010.02.012] [PMID: 20381150]

[27]    Greenlee AR, Arbuckle TE, Chyou PH. Risk factors for female infertility in an agricultural region. Epidemiology 2003; 14(4): 429-36.
[http://dx.doi.org/10.1097/01.EDE.0000071407.15670.aa] [PMID: 12843768]

[28]    Winchester PD, Huskins J, Ying J. Agrichemicals in surface water and birth defects in the United States. Acta Paediatr 2009; 98(4): 664-9.
[http://dx.doi.org/10.1111/j.1651-2227.2008.01207.x] [PMID: 19183116]

[29]    Mesnage R, Clair E, Spiroux de Vendômois J, Séralini GE. Two cases of birth defects overlapping Stratton-Parker syndrome after multiple pesticide exposure. Occup Environ Med 2010; 67(5): 359-9.
[http://dx.doi.org/10.1136/oem.2009.052969] [PMID: 19951932]

[30]    Xavier R Jr, Rekha K, Bairy K. Health perspective of pesticide exposure and dietary management. Malays J Nutr 2004; 10(1): 39-51.
[PMID: 22691747]

[31]    Chakraborty S, Mukherjee S, Roychoudhury S, Siddique S, Lahiri T, Ray MR. Chronic exposures to cholinesterase-inhibiting pesticides adversely affect respiratory health of agricultural workers in India. J Occup Health 2009; 51(6): 488-97.
[http://dx.doi.org/10.1539/joh.L9070] [PMID: 19851039]

[32]    Hoppin JA, Umbach DM, London SJ, *et al*. Pesticide use and adult-onset asthma among male farmers in the Agricultural Health Study. Eur Respir J 2009; 34(6): 1296-303.
[http://dx.doi.org/10.1183/09031936.00005509] [PMID: 19541724]

[33]    Turusov V, Rakitsky V, Tomatis L. Dichlorodiphenyltrichloroethane (DDT): ubiquity, persistence, and risks. Environ Health Perspect 2002; 110(2): 125-8.
[http://dx.doi.org/10.1289/ehp.02110125] [PMID: 11836138]

[34]    Kohler HR. Wildlife Ecotoxicology of Pesticides: Can We Track Effects to the Population Level and Beyond,Science, 2013; 341(6147): 759-65.

[35]    Galloway TS, Depledge MH. Immunotoxicity in invertebrates: measurement and ecotoxicological relevance. Ecotoxicology 2001; 10(1): 5-23.
[http://dx.doi.org/10.1023/A:1008939520263] [PMID: 11227817]

[36]    Rattner BA. History of wildlife toxicology. Ecotoxicology 2009; 18(7): 773-83.
[http://dx.doi.org/10.1007/s10646-009-0354-x] [PMID: 19533341]

[37]    Fleischli MA, Franson JC, Thomas NJ, Finley DL, Riley W Jr. Avian mortality events in the United States caused by anticholinesterase pesticides: a retrospective summary of National Wildlife Health Center records from 1980 to 2000. Arch Environ Contam Toxicol 2004; 46(4): 542-50.
[http://dx.doi.org/10.1007/s00244-003-3065-y] [PMID: 15253053]

[38]    Dzugan SA, Rozakis GW, Dzugan KS, *et al.* Correction of steroidopenia as a new method of hypercholesterolemia treatment. Neuroendocrinol Lett 2011; 32(1): 77-81.
[PMID: 21407165]

[39]    Lin PC, Lin HJ, Liao YY, Guo HR, Chen KT. Acute poisoning with neonicotinoid insecticides: a case report and literature review. Basic Clin Pharmacol Toxicol 2013; 112(4): 282-6.
[http://dx.doi.org/10.1111/bcpt.12027] [PMID: 23078648]

[40]    Gill RJ, Ramos-Rodriguez O, Raine NE. Combined pesticide exposure severely affects individual- and colony-level traits in bees. Nature 2012; 491(7422): 105-8.
[http://dx.doi.org/10.1038/nature11585] [PMID: 23086150]

[41]    Henry M, Béguin M, Requier F, *et al.* A common pesticide decreases foraging success and survival in honey bees. Science 2012; 336(6079): 348-50.
[http://dx.doi.org/10.1126/science.1215039] [PMID: 22461498]

[42]    Cresswell JE, Thompson HM. Comment on "A common pesticide decreases foraging success and survival in honey bees". Science 2012; 337(6101): 1453.
[http://dx.doi.org/10.1126/science.1224618] [PMID: 22997307]

[43]    Rockets, R. Down On the Farm, Yields, Nutrients and soil quality. In: Fox, JE, J. Gulledge, E Engelhaupt, ME Burow and JA McLachlan, Eds. Pesticides reduce symbiotic efficiency of nitrogen-fixing rhizobia and host plants, Proceedings of the national academy of sciences, USA.

[44]    Haefeker, Walter (2000-08-12). "Betrayed and sold out – German bee monitoring". 2007.10-10.

[45]    Zeissloff, Eric (2001). "Schadetimidacloprid den bienen" (in German). Retrieved 2007-10-10.

[46]    Cornell University. Pesticides in the environmentArchived 2009-06-05 at the Wayback Machine. Pesticide fact sheets and tutorial, Pesticide Safety Education Program. Retrieved on 2007-10-11.

[47]    Walley F, Taylor A. Herbicide effects on pulse crop nodulation and nitrogen fixation, FarmTech 2006 Proceedings, 2006; 121-123.

[48]    Hussain S, Siddique T, Saleem M, Arshad M, Khalid A. Chapter 5: Impact of Pesticides on Soil Microbial Diversity, Enzymes, and Biochemical Reactions. Adv Agron 2009; 102: 159-200.
[http://dx.doi.org/10.1016/S0065-2113(09)01005-0]

[49]    Helfrich LA, Weigmann DL, Hipkins P, Stinson ER. (June 1996), Pesticides and aquatic animals: A guide to reducing impacts on aquatic systems. Virginia Cooperative Extension. Retrieved on 2007-1--14.

[50]    Toughill K. The summer the rivers died: Toxic runoff from potato farms is poisoning P.E.I.Archived January 18, 2008, at the Wayback Machine Originally published in Toronto Star Atlantic Canada Bureau. Retrieved on September 17, 2007.

[51]    Pesticide Action Network North America (June 4, 1999), Pesticides threaten birds and fish in California. PANUPS. Retrieved on 2007.

[52]    Singh B, Mandal K. Environmental impact of pesticides belonging to newer chemistry. Integrated pest management,. Jodhpur, India: Scientific Publishers, 2013; pp. 152-90.

[53]    Trajkovska S, Mbaye M, Gaye Seye MD, Aaron JJ, Chevreuil M, Blanchoud H. Toxicological study of pesticides in air and precipitations of Paris by means of a bioluminescence method. Anal Bioanal Chem 2009; 394(4): 1099-106.
[http://dx.doi.org/10.1007/s00216-009-2783-z] [PMID: 19387620]

[54]    Hussain S, Siddique T, Saleem M, Arshad M, Khalid A. Impact of pesticides on soil microbial diversity, enzymes, and biochemical reactions. Adv Agron 2009; 102: 159-200.
[http://dx.doi.org/10.1016/S0065-2113(09)01005-0]

[55]    RuizRomera E, Antiguedad I, Garbisu C. Tebuconazole application decreases soil microbial biomass and activity. Soil Biol Biochem 2011; 43: 2176-83.
[http://dx.doi.org/10.1016/j.soilbio.2011.07.001]

[56]    Ware GW. Effects of pesticides on nontarget organisms. Residue Rev 1980; 76: 173-201.
[http://dx.doi.org/10.1007/978-1-4612-6107-0_9] [PMID: 7006022]

[57]    Kevan PG. Pollinators as bio-indicators of the state of the environments: Species, activity and diversity. Agric Ecosyst Environ 1999; 74: 373-93.
[http://dx.doi.org/10.1016/S0167-8809(99)00044-4]

[58]    Fishel FM, Ferrell JA. Managing pesticide drift Agronomy department PI232. Gainesville, FL, USA: University of Florida 2013.

# CHAPTER 8

# Water Pollution

**T. Dhanalakshimi**[1,*] and **M. Sudha Devi**[2]

[1] *Department of Biology, Ministry of Education, Republic of Maldives*

[2] *Department of Biochemistry, Biotechnology and Bioinformatics, School of Biosciences, Avinashilingam Institute for Home Science and Higher Education for Women, Coimbatore 641 043, Tamil Nadu, India*

**Abstract:** Water is indispensable for life, and is indisputably the most priceless natural resource that exists on our planet. Living things on the planet cannot live without water. Though we as humans know this fact, we neglect it by polluting our natural resources like rivers, lakes, oceans, and the water beneath the surface of the earth, groundwater. Consequently, we are harming our planet to the point where organisms are vanishing at a very alarming rate. Most of the organisms die, and our drinking water has been prominently affected, as we use water for amusing purposes. It then affects the climate, resulting in consequences such as drought, changes in weather patterns, etc., which have ultimately increased the demand for freshwater. In order to combat water pollution, we must understand the problems and become part of the solution. The change in the chemical and physical properties of water is called water pollution and thus directly or indirectly harmfully affects the living organisms that consume it and render it unfit for the required uses. The water gets polluted through various sources of the organic and inorganic pollutants of Industry, Agriculture, Domestic, Thermal and Biological wastes. Water pollution is a major global problem thatrequires ongoing evaluation and revision of water resource policy at all levels. It has been suggested that it accounts for the deaths of more than 14,000 people daily. In addition to the acute problems of water pollution in developing countries, developed countries continue to struggle with pollution problems as well. In this paper, the sources of water pollution, effects of water pollution on the ecosystem, ways to control pollution and conservation have been described.

**Keywords:** Classification, Health and Environmental Concern, Pollution, Resources, Water.

## INTRODUCTION

Pollution is the introduction of contaminatants into the natural environment by human activities and causes environmental pollution. The term *pollution* refers

---

* **Corresponding author T. Dhanalakshimi:** Department of Biology, Ministry of Education, Republic of Maldives; Tel: 9965664834; E-mail: dhanvarshar@gmail.com

**J. Senthil Kumar, P. Ponmurugan & A. Vinothkanna (Eds.)**

primarily to the entangling of air, water, and land, and is a much more complicated global problem. We, as human beings, are forced to live with contaminated air, water and all sorts of distressing noises. Air and water pollutions are major fatal pollution and can cause the death of many organisms in the ecosystem, including humans. The acute problem in developing and developed countries is water pollution. The damping of an enormous amount of harmful or insufferable matter in water to alter the natural quality of water is called water pollution.

**Water pollution** is the change in the qualities of the water by the presence of surplus physical, chemical or biological substances, which can change the taste, smell and transparent nature of water and is capable of causing harmful effects to the living organisms. The natural or pure water is colorless, odorless and transparent. However, some water pollutants could not be seen or tasted by naked eyes, for example, some chemicals, such as pesticides, and most of the microorganisms that cause water-borne diseases. Contaminated water should not be used for drinking, washing, bathing or agriculture. If polluted water is used by humans, then it can adversely affect the body in different ways, depending on the type and concentration of pollutants [1].

If these pollutants are not properly disposed or filtered, they might spread throughout the water and affect all living organisms that come into contact with it. In addition to harming animals and humans, it can also affect plants, trees, the soil and other natural things and an earth resource. This article provides with an in-depth explanation of water pollution, the source and effects associates with it, and the prevention of water pollution.

## TYPES OF POLLUTION

These main categories of common water pollutants which effect the environments are:

1. Organic pollutants
2. Inorganic pollutants
3. Water borne pathogens
4. Nutrients
5. Agricultural runoffs
6. Suspended solids and sediments
7. Thermal pollutants
8. Radioactive pollutants.

According to the polluting substances, environmental conditions and living things involved, influence pollutants in natural waters disparities [1].

## SOURCES OF WATER POLLUTION

Several things pollute the water resources like river, lack, pond, sea and groundwater by the man-made activities (Fig. **1**) [2] and by natural disasters.

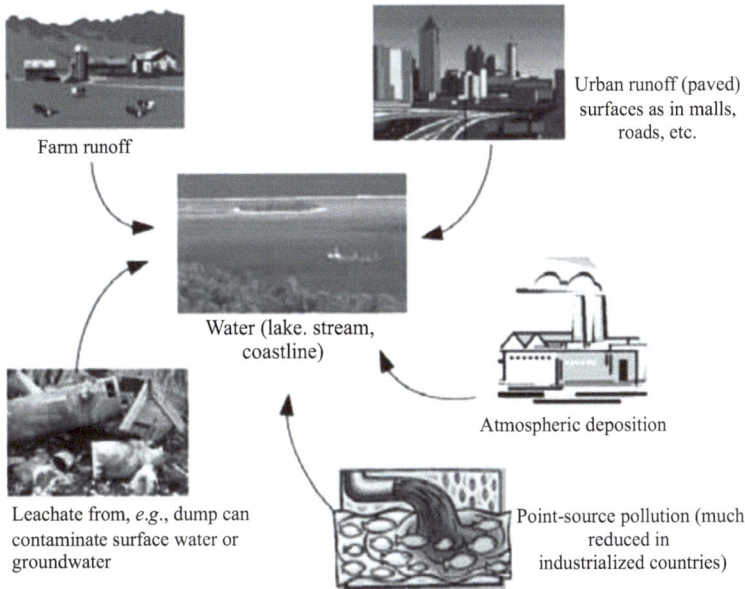

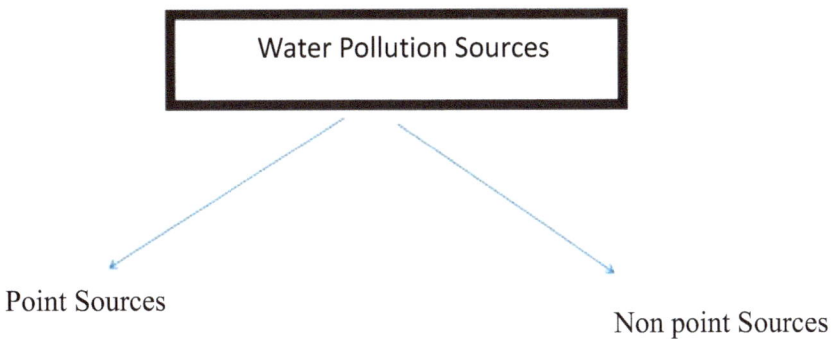

Farm runoff

Urban runoff (paved) surfaces as in malls, roads, etc.

Water (lake. stream, coastline)

Atmospheric deposition

Leachate from, *e.g.*, dump can contaminate surface water or groundwater

Point-source pollution (much reduced in industrialized countries)

Water Pollution Sources

Point Sources

Non point Sources

Fig. (1). Sources of Water Pollution [2].

In the greatest cases, water pollution is caused due to man-made activities and can be categorized based on the source and characteristics such as point and non-point source pollution.

Point-source pollution can be defined as pollution that comes from direct contact with a single point of pollutant discharge from domestic, sewage treatment plants, power plants, factories, ships, injection wells, and some manufacturing or industrial sources. Non-point sources pollution can be defined as pollution that comes from many diverse or diffused sources rather than specific identifiable and specific points like mining activities, agricultural and urban runoffs. Non-point sources pollution is the main account for the greater part of contaminants in water sources [3].

## Industrial Effluent

Contamination of natural water resources from industries is a result of various types of industrial processes and disposing of various wastes. After the Industrial Revolution, drastic growth in the social and economics of humans on one side of development and another side is the mischievous effect of industrial pollution. Moreover, Industries use a huge quantity of water for the processing of materials, and these polluted water discharges into the water sources without any proper treatment. This kind of activity causes water contamination and is considered great hazard to our environment which affects not only the quality of soil, crop and environment but also dangerous to the all living things, however, which degrades the quality of food crops [4].

Industrialization is a major step in the development and urbanization of society. Although there are merits and demerits, it has been revealed that it is the foremost threat to the environment as it discharges different toxic chemicals, gases, solid wastes, as well as microbes of different kinds into our environments such as land, air, and water. However, it mainly affects water resources and causes different dangers to all living things. [5, 6].

## The Basis of Industrial Water Pollution

A lack of stern government policies in developing countries.

The uses of old-fashioned technologies, which cause more pollutants than advanced technologies.

More pollution because of unplanned industrial growth.

Due to the lack of capital, they are leading the industry without taking any pretreatment process.

The release of industrial effluents or wastewater into contiguous terrestrial or aquatic habitat has been concerned as one of the major sources of environmental pollution being the by-products of human or industrial activities [ 7 ].

## Municipal and Domestic Wastes

The various toxic substances present in untreated or improperly treated wastewater effluents are fetal to all the living things in the environment. The nutrients are major contaminants in wastewater effluents such as nitrogen and phosphorus, heavy metals, organic compounds, hydrocarbons, microbes and endocrine disruptors, and which leads to undesirable effects on human health and the environment [8].

Municipal wastes are disposing by landfills on the environment (Fig. **2**) and recognized as one of the major threats and contaminants to groundwater resources. The greater possibility of groundwater contamination takes place at nearby landfills because of the potential pollutants leachate [9 - 15].

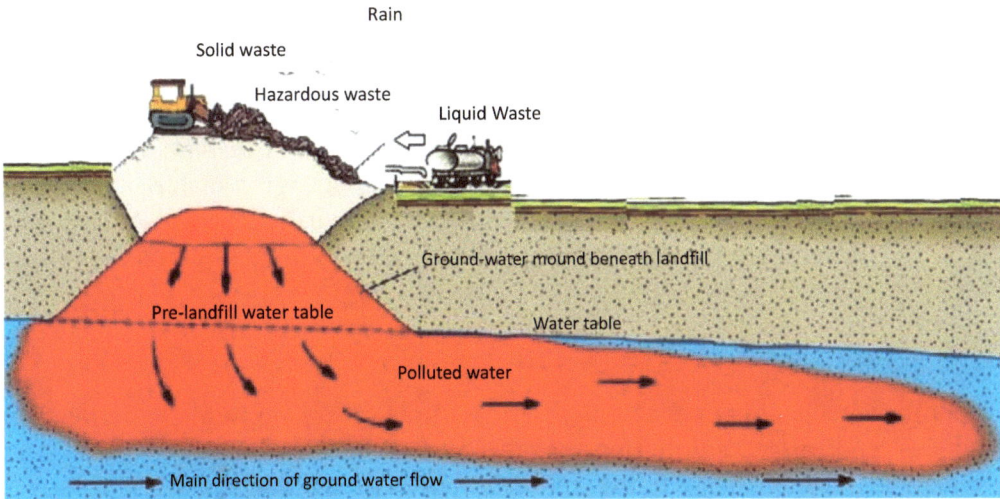

**Fig. (2).** Landfilling practice [9].

This kind of contamination of groundwater resources is a significant risk to local resource users and the natural environment [6] which is difficult to retain the original quality. In developing countries, the groundwater quality has been deteriorating by point and non-point source, including domestic and municipal wastewater uses, and has become an increasing problem in the recent year [9]. Even if it comes to unlimited landfills without any lining system and without checking the kind of waste received, landfill leachate contains various types of contaminants, including heavy metals like Cadmium, Nickel, Zinc, Copper and Lead, and these metals are found at restrained concentration levels in municipal wastes.

It is very important to perfectly design and builds facilities to put off the impact of the landfill emissions on quality of ground water and environment. In order to maintain the quantity and quality of leachate generation and the probable effects on groundwater, quality must be recommended to regulate reliable long term monitoring over the period of time. However, it is important to educate all levels of society as well as civil society and stakeholders on the significance of groundwater and the probable outcomes from the mismanagement of solid waste and landfilling habits in developing countries and should encourage the participants [16].

Toilets, latrines, wastewater from kitchens and bathrooms are the sources of domestic pollution; however, these wastes contain various pollutants of chemicals like detergents and some pathogens. These cause pollution and affect the environment if the wastes are not properly treated and get into the environment, which can be washed into water resources like river and other sources. Some of the organic wastes are also obtained from domestic waste sources, including excreta, food waste and other kitchen waste like cooking oil residues. Though many wastes from shops, households, markets, hotels and restaurants and some businesses, including fruits and vegetables, wastes, packaging materials and other types of garbage, are considered as domestic wastes. Air pollution also happens in the form of smoking fuels and fires and releases a huge level of carbon dioxide to the environment and causes pollution.

**Agricultural Waste**

Inappropriate water management and irrigation of agricultural operations and practices lead to water pollution by runoff of surface, pollute both surface and groundwater. The plants, wildlife, humans, animals and aquatic life are negatively affected because of the use of fertilizers, pesticides, herbicides, manure and agrochemicals, and it leads to contamination of waterways, groundwaters and deflates water quality [17].

Agriculture is important in developed and developing countries that promote economic growth. However, agricultural activities might have harmful impacts on the environment and can be defined as a non-point source of pollution, which causes mainly water pollution *via* soil [18]. In spite of the uses of chemical fertilizers in agriculture, which are the main sources of water pollution and affect severely water systems, groundwater, aquatic life forms. Though, the agrochemicals, pesticides and fertilizers are absorbed by the aquatic living organisms and affecting their health especially, their reproductive system. The main sources of agricultural wastes are huge levels of nitrogen, phosphorus from fertilizers and animal wastes, and these flow into the surface of waters, which depletes the dissolved oxygen, increases the BOD levels, and eradicates the aquatic lives by the eutrophication. These fertilizers have some chemical compounds such as heavy metals and ammonia; those can become lethal to aquatic lives and humans who consume these fishes [19].

Nowadays, formers aware of fatal effects of the nitrogenous fertilizers on the environment, these can reach the water environment by the following ways:

1. Drainage
2. Leaching
3. Flow

Nitrate leaching is also particularly connected to agricultural practices of fertilizing and cultivation, and thus, huge amounts of nitrate accumulate in the soil and along with water evaporation in some of the irrigated agricultural land of arid and semiarid regions.

Based upon the circumstance, an accumulation of nitrate is leached in various levels and reaches the depth of soil. Fertilizers are converted into nitrate through nitrification by microbes. Since nitrates get a negative charge and reach groundwater. But still, in perfect state, the plants use 50% of nitrogenous fertilizers implemented on soil, 15- 25% react organic compounds in the clay soil, 2-20% lost by evaporation, 2-10% hold up in surface and groundwater [20, 21]. The maximum of nitrogenous fertilizers are not taken up products, and they interfere with surface and groundwater.

Another main reason is the use of chemical fertilizers in agriculture. It results in many environmental problems that fertilizers contain heavy metals such as cadmium and chromium and a large concentration of radionuclides [22]. The accumulation of inorganic pollutants has occured because of heavy metals and radionuclides in plants [23].

During the season, the huge amounts of chemical fertilizers used and affect aquaculture, which causes greenhouses, particularly the fertilizers pollute the well water and water resources and produce deteriorates levels of crop production in quantity and quality-wise. Thus, the reason for all these problems is the usage of a high level of fertilizers, which contains an excess amount of nitrate. As well as, the phosphate might increase in drinking water and rivers as a result of the flow of phosphorus from the surface of the soil. The chemical fertilizer contains carcinogenic substances such as nitrosamines because of the harmful accumulation of $NO_3$ and $NO_2$ [23, 24].

## Spilling of Oil

Oil pollution defines as undesirable polluting effects that oil spills have on the environment and living things, including humans, because of the environmental liberation of different organic compounds that made up crude oil and distilled products and including various hydrocarbons. The carbon and hydrogen atoms bind together forms hydrocarbons, which resulting in paraffins (Alkanes), isoparaffins (isoalkanes), aromatics (Benzene or various PAHs), cycloalkanes and unsaturated alkanes9lkanes and akkynes). Some other than these components also discharge such as sulfur, nitrogen and oxygen atoms [25].

Water gets polluted from oil spills, routine shipping, run-offs and dumping on a daily basis and about 12% of the oil enters into the ocean by spilling of oil; the reaming come from shipping travel, drains and dumping. One of the severe problems is the spilling of oil from a tanker into one place. Spilling of oil causes a localised problem, and it can be terrible for marine and wildlife such as fish, birds and sea otters. The oil does not dissolve in water and forms a dense coagulant in the water, like stifle fish, the feather of marine birds gets wedged and they could not fly and blocks light from photosynthetic aquatic plants.

The second most frequent kind of pollutant on land water is oil and fuels. There are different preventive measures in place to deal with oil pollutions, including mineral oils, fuel oils and vegetable oils, and can be identified as possible for further action. The physical property of oil is highly visible, which affects the environment in a number of ways, such as reduce the dissolved oxygen. This property affects drinking water, making it unfit for other uses.

According to the Groundwater Regulations, it is unlawful to release mineral oil into groundwater because mineral oil is a hazardous substance. It might be complicated to deal with the contamination of groundwater with oil. The long-term effects are polluting surface water and drinking water supplies and harmingwildlife also. Wildfowl are especially vulnerable, both through harm to the waterproofing of their plumage and through swallowing oil during the groom. A water vole is one of the mammal variety; it is also affected and not good to consume when fish is exposed to oil. Oil is used in large quantities everywhere in society, need a wide range of distribution and storage system. There may be probable spills, other accidental releases, and loss from storage during delivery or distribution, and these should not dispose to the drainage system because it is illegal (Fig. **3**) [26].

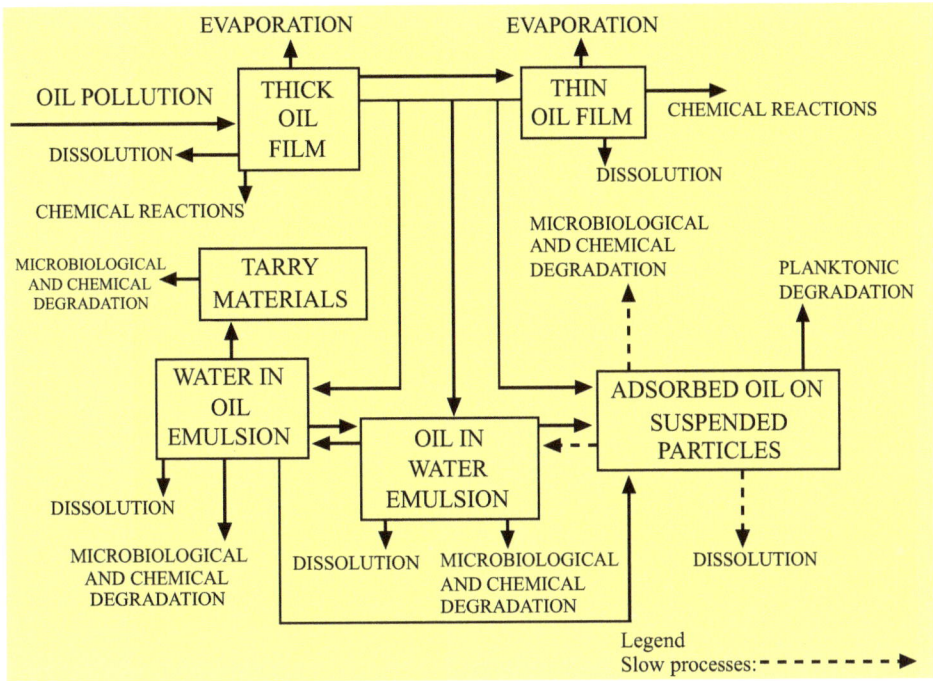

**Fig. (3).** Disposal of waste oil to drainage systems [26].

A sea otter feeds on a sea urchin at the water's surface in Alaska's Kachemak Bay National Estuarine Research Reserve (NOAA).

The removal of oil from the water surface by skimmers, which glide over the water and comb-out the oil is being implemented, and oil-water mix can be removed.

Use the floating boom as a part of the effort to stop oil from reaching the coast and made up by using the nylon tights, animal fur and human hairs. Since hair donation helps to make the special boom and laid on beaches to immerse up any oil the wash ashore.

For breaking down the oil, dispersant chemicals have been sprayed from ships and aircraft to degrade by wind and waves.

Another tricky method is burning, but it can be associated with environmental risk like release toxic smoke [27].

## MINING WASTE

Mining is another kind of water pollution, which includes acid mine drainage and metal contamination, which increases the sediment levels in water streams. The main sources are active and huge metal contamination on the surface and underground mines, processing plants, waste disposal area, transporting roads and tailing ponds.

Though, mining influences freshwater by using the huge amount of water in ore processing and discharging mine effluent and leakage from tailings and waste rock impoundments.

Exponentially, man-made activities include mining that threatens the water resources on which human beings depend. Thus, water has been called mining's quite common casualty [28].

They percolate awareness of the environmental legacy of mining activities with small concern about the environment.

The following are the four important types of mining impacts on the water quality.

### Heavy Metal Contamination

Metals such as arsenic, cobalt, copper, cadmium, lead, silver and zinc present in excavated rock or exposed in an underground mine get into the water. These metals are leached out and taken to downstream as water washes over the rock surface. Since metals can leach at low pH conditions, mobile at neutral pH, and leaching formed by acid mine drainage.

### *Acid Mine Drainage*

The rock contains suphide and when it is exposed to air and water to become sulphuric acid is called Acid Mine Drainage by a natural process. It harshly degrades water quality, lethal to aquatic life, and makes almost unusable due to the presence of sulphuric acid.

### *Processing Chemicals Pollution*

Some of the chemicals such as sulphuric acid and cyanide are highly toxic to the animals, which are used to separate the target mineral from the ore, and these chemicals spill, leak or leach from the mine location into nearby water bodies.

## *Erosion and Sedimentation*

Due to the building construction, road sanitation, open pits and waste impoundments, soil and rocks get disturbed, which get into the water streams. This disproportionately sediment can block riverbeds and smother watershed, wildlife habitat and aquatic life.

Recently, there is an improvement in mining practice, and even there is an environmental risk stay. One of the issues is that mining has become more mechanized and can handle many rocks and ore material than ever before. Therefore, mine wastes are multiplied abundantly. Advanced mine technologies are developed to make gainful to mine less level grade ore, and more waste will be produced in the future [29, 30].

### Power Plant and Radioactive Pollutants

The discharging of pollutants to the surface water from power plants is badly effecting water. The usage of "Once-Through Cooling", cooling water is taken from the lake and river and used to condense the steam for circulating through the plant. The used water pick up heat from the steam and is returned to the same lake and river at a hot temperature. Effect of this technology, the warming of the lake or river near the released point, harmfully affects the temperature-sensitive plants, aquatic life, and microbial activity, and there are some chemical and physical reactions in the water. Even if we are using cooling towers instead of once-through cooling also discharges the process water and which is warmer than ambient water temperatures can change the local fishery composition, aquatic macroinvertebrate communities and aquatic plants, and some diseases and fish contamination occurs because of the surface water pollutants.

For generating electricity, burns the fossil fuel, biomass to make either hot air or steam necessary to spin power turbines, and the same in nuclear power plants also use it for a nuclear reaction, Because of that, creates exhaust gases and other by products which cause air pollution even. Utilize huge quantity of water to make steam for these plants from nearby river and lakes or from underground water, and it should be purified before discharged again into same water sources by proper treatments based on the concentration and temperature of the pollutants and water, respectively.

A large dose of radionuclides exist radioactive materials in water sources to which the living systems are adopted with different negative effects.

The sources of radioactive pollutants are nuclear power plants, nuclear reactors, nuclear tests, nuclear installations, operations of power, fission processes and fusion products, through which water streams get pollute. These radioactive molecules deposit in the living bodies and organs deliver radiation dose in hazardous level, and extreme toxic radioactive elements such as Pu, Np, Cm, Bk, Cs and Ruetc, arereleased from the neutron bombardment of atomic fuel. Once it enters the water bodies, interrupts the ecocyling process, enters into the food chain, and ultimately effects metabolic pathways. The fallout of radioactive materials from explosions of nuclear weapons is the main concern, besides that hazardous associated with it. The other source of pollution is the accumulating of radioactive waste materials from nuclear plants, nuclear reactors and waste from medical and research laboratories [31].

Power plants should take proper initiative to filter and purify the water before discharging into the water sources and dispose of solid by-products appropriately. The used water should be cleaned, filtered, and processed by neutralizing both temperature and any dissolved materials as soon as before discharge into the water resources. Some microelements such as mercury may retain in the power plant discharge water even after treatment; this can be regulated by DNR *via* WPDES(Wisconsin Pollution Discharge Elimination System) permits and should have been limited and monitored by them. For a description of the WPDES program, go to the DNR [32].

**Biological Waste**

Water sources get contaminated by physical, chemical and microbial populations and affect the environment and humans. The microbial populations deteriorate and cause a different water-related disease called water-borne pathogens. Water contains a broad range of biological organisms, which are called biological pollutants and get into the water in various ways such as natural water organisms, soil microbes, intestinal bacteria, sewage bacteria, and pathogens. Biological pollutants contain water-borne pathogens such as bacteria, fungi, algae, viruses, pathogenic protozoa, parasite worms, helminthic parasites and indeed of any animals and plants that multiply extremely in the water bodies for the other reason. This may be occurring as a result of mainly domestic sewage, human excreta and from some industries [33]. These biological pollutants are responsible for water-borne diseases such as typhoid fever, cholera, dysentery, polio, hepatitis, and schistosomiasis. Coliform bacteria are an indicator organism to find water quality.

The sources of microbial contamination into rivers and lakes, are wild type and domestic animals and humans, among which human sources are dangerous because of the spreading pathogens from person to person [34]. Another equally harmful source is animals, and some animals carry certain diseases, causing pathogens, for example, waterfowl, which carries the influenza virus [35]. Some domestic purpose water of river water polluted by microbes is hazardous to humans health [36]. The main source of drinking water is the river, and it has microbial pathogens, which would cause diarrhoea only to under- five-year children in developing countries, which is the third foremost cause of death [37, 38].

Though advanced technology exists and is used in the public water system and some rural system, nowadays, available technologies eliminate diseases causing constituents like waterborne pathogens, remove all naturally present and artificial synthetic organic chemicals, limit some metals and radioactive elements, and capable to effectively protect human health as well as maintain the reliability of the distribution system. Although advanced technology does exist, it is infrequently used on most of the public water systems and rarely on the smaller rural systems. Nevertheless, a state-of-the-art treatment plant that used all available technologies to remove disease-causing constituents such as bacteria and viruses removed all naturally occurring and synthetic organic chemicals, limited the amounts of naturally occurring metals and radioactive elements, and could effectively protect human health as well as maintain the integrity of the distribution system [39].

### Hazardous to Our Earth

Water gets polluted in different ways, and they all have deteriorating effects on the environment, which affects the earth. Once an environment gets polluted, it sequentially affects the green and pure earth. If any of the earth's natural resources (air, soil or water) gets polluted, it will sequentially spread to all the sources. Air and water are foremost important for all living things. Hence, water pollution has endangered many living things, including human and aquatic animals, so we, as biologists, need to take some remedial measures if not to get rid of it, but at least to diminish the water pollution. We can control the domestic, industrial and biological waste and should do a proper treatment before discharging into the natural water sources.

### CONCLUSION

Water is indispensable to all on earth; a condition of sustainable growth should be to make sure streams, rivers, lakes and oceans without contamination. The problems related to water pollution have the potential to interrupt life on our earth to an immense extent. The legislative body has passed acts to try to combat water pollution, thus recognizing the fact that water pollution is, indeed, a grim issue. But the government alone cannot solve the entire problem, and we, as individual people, should take responsibility and be involved to face the water issues. We must educate either ourselves or some social representatives about the local water resources and learn about ideas for disposing of harmful wastes. Therefore, it should not end up as wastes without any pretreatment or recycle process. In the 21$^{st}$ century, awareness and education are the most important weapons to deal with pollution and control. Unless this problem continues, lives on earth suffer relentlessly. Hence, when all the people from the government, politics, public and youth, work together, only then could solve water and all kind of problems, and could protect our planet.

## CONSENT FOR PUBLICATION

Not applicable.

## CONFLICT OF INTEREST

The authors confirm that this chapter contents have no conflict of interest.

## ACKNOWLEDGEMENTS

Declared none.

## REFERENCES

[1]     Benson I, Pollution W. Water Pollution: Encyclopedia of Global Warming and Climate Change. 2008; 1 – 3: pp. 183-8.

[2]     Marquita K. Hill, Understanding Environmental Pollution. 3$^{rd}$ed ., Cambridge University Press 2010. ISBN-13 978-0-511-90782.

[3]     Monty C. What is water Pollution? , Texas cooperative Extension, SCS 2005-02, Water and Me series. http://soilcrop.tamu.eduhttp://water.tamu.eduhttp://waterandme.tamu.edu

[4]     Anwar Hossain M, Khabir Uddin M, Molla A H, Afrad M S I. Anwar Hossain M, Khabir Uddin M, Molla AH, Afrad MSI, Rahman MM, Rahman GKMM. Impact of Industrial Effluents Discharges on Degradation of Natural Resources and Threat to Food Security. The Agriculturists 2010; 8(2): 80-7.

[5]     Inyinbor AA, Adekola FA, Olatunji GA. Liquid phase adsorption of Rhodamine B onto acid treated Raphiahookerieepicarp: Kinetics, isotherm and thermodynamics studies. S Afr J Chem 2016; 69: 218-26.
        [http://dx.doi.org/10.17159/0379-4350/2016/v69a28]

[6]     Rana RS, Singh P, Kandari V, Singh R, Dobhal R, Gupta S. A review on characterization and bioremediation of pharmaceutical industries' wastewater: An Indian perspective. Appl Water Sci 2017; 7: 1-12.
        [http://dx.doi.org/10.1007/s13201-014-0225-3]

[7]     Odutayo FOI, Oyetade OB, Esan EB, Adaramola FB. Effects of Industrial Effluent on the Environment Using Allium CepaAnd Zea Mays as Bioindicators. Int J Environ Pollut Res 2016; 4(4): 1-12.

[8]     Davies P S. The biological basis of wastewater treatment. West of Scotland: Strathkelv in instruments Ltd 2005.

[9]     Aljaradin M, Persson KM. Persson, municipal landfilling practice and its impact on the water resources – jordan. World Environment 2014; 4(5): 213-8.

[10]   Aljaradin M, Persson KM. Environmental impact of municipal solid waste landfills in semi-arid climates - case study – jordan. Open Waste Manage J 2012; 5: 28-39.
[http://dx.doi.org/10.2174/1876400201205010028]

[11]   Nixon WJ. Murphyr, and R. Stessel, An empirical approach to the performance assessment of solid waste landfills. Sage 1997; 15.

[12]   Shiraiwa SS, and Silvino A. Evaluation on surface water quality of the influence area of the sanitary landfill. EngenhariaAmbiental Pesqui Tecnol 2008; 5: 139-51.

[13]   Aljaradin MNS, Persson KM. A study of the groundwater quality in the surroundings of the mafraq landfill, Jordan. Journal of Water Management and Research (Vatten) 2012; 68: 97-101.

[14]   Aljaradin M, Persson KM. Proposed water balance equation for municipal solid waste landfills in Jordan. Waste Manag Res 2013; 31(10): 1028-34.
[http://dx.doi.org/10.1177/0734242X13492003] [PMID: 23797298]

[15]   RSS and M. Env. Studying selected landfills in Jordan and assessing nearby groundwater quality. 2008.

[16]   Mor S, Ravindra K, Dahiya RP, Chandra A. Leachate characterization and assessment of groundwater pollution near municipal solid waste landfill site. Environ Monit Assess 2006; 118(1-3): 435-56.
[http://dx.doi.org/10.1007/s10661-006-1505-7] [PMID: 16897556]

[17]   https://www.quora.com/How-do-agricultural-wastes-harm-the-environment

[18]   Semaan J, Flichman G, Scardigno A, Steduto P. Analysis of nitrate pollution control policies in the irrigated agriculture. Agric Syst 2007; 94: 357-67.
[http://dx.doi.org/10.1016/j.agsy.2006.10.003]

[19]   https://www.quora.com/How-do-agricultural-wastes-harm-the-environmen

[20]   Korkmaz K. TarımGirdiSistemindeAzotveAzotKirliliği. 2007. http://www.ziraat.ktu.edu.tr/tarim_girdi.htm

[21]   Sönmez M. Kaplan S. Kimyasalgübrelerinçevrekirliliğiüzerineetkileriveçözümönerileri. BatıAkdeniz TarımsalAraştırmaEnstitüsüDerimDergisi 2008; 25(2): 24-34.

[22]   Serpil S. An Agricultural Pollutant: Chemical Fertilizer. Int J Environ Sci Dev 2012; 3: 1.

[23]   Sönmez I M. An investigation of seasonal changes in nitrate contents of soils and irrigation waters in greenhouses located in antalya-demre region. Asian Journal Of Chemistry 2007; 19(7): 5639-46.

[24]   Çevreve TC. OrmanBakanlığıTürkiyeÇevreAtlası ÇED PlanlamaGenelMüdürlüğüÇevreEnvanteri DairesiBaşkanlığı. Ankara 2004.

[25]   https://response.restoration.noaa.gov/oil-and-chemical-spills/ oil-spills/ oil-spills-water-surface.html

[26]   Csaba P, Csaba J Water resources management and water quality protection ,DebreceniEgyetem a TÁMOP 412 pályázatkereteinbelül 2011.

[27]   https://response.restoration.noaa.gov/oil-and-chemical-spills/ oil-spills/ oil-spills-water-surface.html

[28]   de Rosa C, Lyon J Interview, mineral policy center, Washington DC https://www.safewater.org/ fact-sheets-1/2017/1/23/miningandwaterpollution

[29]   https://www.safewaer.org/fact-sheet-1.2017/ 1/23/ miningandwaterpollution

[30]   de Rosa C, Lyon J. Golden Dreams, Poisoned Streams. Washington, DC: Mineral Policy Center 1997.

[31]   Sharma BK. Water Pollution. 4th ed. Meerut: Goel Publishing House 2005; p. 85.

[32]   http://dnr.wi.gov/topic/wastewater
https://psc.wi.gov/Documents/Brochures/Enviromental%20Impacts%20of%20PP.pdf
http://psc.wi.gov/

[33]   Kumwenda S. Tsakama M, Kalulu K, Kambala C. Determination of biological, physical and chemical pollutants in mudi river, blantyre, malawi. J Basic Appl Sci Res 2012; 2(7): 6833-9.

[34]   Hadipour MM. Seroprevalence of H9N2 avian influenza virus in human population in boushehr province, Iran. Asian J Anim Vet Adv 2011; 6(2): 196-20.
[http://dx.doi.org/10.3923/ajava.2011.196.200]

[35]   Becker AM, Gerstmann S, Frank H. Perfluorooctane surfactants in waste waters, the major source of river pollution. Chemosphere 2008; 72(1): 115-21.
[http://dx.doi.org/10.1016/j.chemosphere.2008.01.009] [PMID: 18291438]

[36]   Burns D. Handbook of water purity and quality [Internet] [cited 2012 Mar 15] 2009.

[37]   Cunliffe NA, Kilgore PE, Bresee JS, *et al.* Epidemiology of rotavirus diarrhoea in Africa: a review to assess the need for rotavirus immunization. Bull World Health Organ 1998; 76(5): 525-37.
[PMID: 9868844]

[38]   WHO. Global Water Supply and Sanitation Assessment. Geneva: World Health Organization 2000.

[39]   Sullivan PJ, Agardy FJ, Clark JJJ. Water Protection. In: The Environmental Science of Drinking Water. Oxford, United Kingdom: butterworth-heinemann 2005: pp. 89-141.
[http://dx.doi.org/10.1016/B978-075067876-6/50006-3]

# CHAPTER 9

# Marine Pollution

**B. Sathya Priya**[1,*], **K. Chitra**[2] **and T. Stalin**[3]

[1] *Department of Environmental Sciences, Bharathiar University, Coimbatore - 641046, Tamil Nadu, India*

[2] *Department of Botany, Bharathiar University, Coimbatore-641046, Tamil Nadu, India*

[3] *Forestry Research and Development Unit, Molecular Biology Division, Karur-639 136, Tamil Nadu, India*

**Abstract:** Marine resources are very important to human beings as they provide seafood, tourism, trade, sports, and livelihoods. The marine pollutants, such as oil spills, marine debris, heavy metals, plastics, toxic chemicals, noise, nutrients, radioactive waste, sewage, and industrial discharge, severely threaten the marine ecosystem, which includes mangroves, coastal wetlands, and other coastal habitats. The marine pollution disintegrates the significant oceanic and mangrove biodiversity at a faster rate. The population explosion, increased industries, and advanced technology are responsible for releasing more toxicants and man-made debris, which severely affect the flora and fauna in the marine ecosystem. This study deals with the sources, impacts, and control measures of marine pollution.

**Keywords:** Heavy Metals, Marine Biodiversity, Marine Debris, Marine Pollution, Micro Plastics, Oil Spills, Plastics.

## INTRODUCTION

The ocean is an important place for us to harvest food, for playing and gaining economy . However, it has been treated as a universal dumping ground for many centuries. Much of our waste products are dumped into the marine ecosystem as a waste receptacle due to the increased population. It was found that 80 - 90% of raw and untreated sewage was discharged into coastal waters from the developing countries. The untreated sewage (90-95%) and untreated effluents (75%), with toxicants and metals, are discharged into the Bay of Bengal. It creates a coastal dead zone that depletes the oxygen level in the water. Marine pollution threatens biodiversity, including flora and fauna, climatic conditions, tourism revenue, economic activities, quality of seawater, and the environment.

* **Corresponding author B. Sathya Priya:** Department of Environmental Sciences, Bharathiar University, Coimbatore - 641046, Tamil Nadu, India; Tel: 8825688512; E-mail: sbspriya11@gmail.com

**J. Senthil Kumar, P. Ponmurugan & A. Vinothkanna (Eds.)**

Trash, such as plastics and other man-made substances, metals, oil spills, dumped waste, energy gain, noise during marine activities, sewage, industrial discharge and agricultural waste, are the important sources of marine pollution. The consumption of highly contaminated seafood induces infectious diseases, such as hepatitis, which increases the health risks and medical costs. The organic pollutants, which are persistent in the marine ecosystem, cause reproductive disorders and cancers. The polycyclic aromatic hydrocarbons (PAHs) cause chromosomal abnormalities and congenital heart disease [1 - 8]. Medical waste contamination leads to chronic health problems in human beings who consumed seafood.

## SOURCES OF MARINE POLLUTION

### Oil Pollution

Oil spills, either crude oil or refined petroleum products due to leakage and oil tanker accidents, cause a severe effect on marine life. Oil inputs in the marine ecosystem primarily originate from petroleum use, petroleum transportation, petroleum extraction, exploration, and production activities.

### Marine Debris

Marine debris is one of the important hazards to sea birds and mammals. Nearly 80% of marine debris comes from land sources and the remaining comes from sea-based sources, such as cargo or fishing boats. The main trash is plastic and includes other waste substances. During rain, litter is moved into storm drains and directly reaches seawater.

### Heavy Metals

Deep-sea mining, ship scrapping, and industrial effluent discharge also lead to heavy metal accumulation in the marine ecosystem.

### Nutrients Enrichment

The ocean is also contaminated due to nutrients, such as livestock waste, detergents from houses, discharge from sewage treatment plants, industrial discharges, septic system leakages, crop and lawn fertilizers, coming out from land during the rainy season. The agricultural fertilizers with high nitrogen and phosphorus content enhance nutrient enrichment.

### Plastic Debris

Plastics are synthetic polymers. They are non-biodegradable, and remain in the environment for many decades and cause serious threats to marine life. Plastic pollution originates mainly from land-based sources (80%), while the remaining comes from ocean-based sources. The improper waste management of landfills, illegal dumping, inadequate disposal of plastic packages, disposal of plastic bags, containers, bottles by beachgoers, storm water discharge with plastic debris, discharge of plastic materials during shipping, boating, and plastic debris released by cruise ships, are the main sources for causing plastic pollution in oceanic region.

## Toxic Chemicals

During the rainy season, toxic chemicals from the agricultural fields, industrial areas, household cleaning, gardening, and personal care may enter the ocean. The residual toxicants remain in the environment for many years after their use. Mercury is released during incineration of medical waste, and fossil fuels as well as from industrial discharge. The inorganic mercury is converted into methyl mercury, which is an easily available toxic form. It is bioaccumulated in marine organisms due to its solubility in fatty tissues and is transferred to higher trophic levels through the food chain. The chemicals may be transformed or reduced due to bacterial metabolic activities. The organic pollutants, such as PCBs, PAHs, and dioxin, which are persistent in marine sediments for decades, also cause a threat to marine diversity. More quantities of oil-based chemicals, such as PAHs (Poly Aromatic Hydrocarbons), and sealants with PCBs (Polychorinated Biphenyls,) are discharged into the coastal region during dismantling and scrapping of ships.

## Sewage and Industrial Discharge

Untreated sewage and spills due to leaking pipes from septic household tanks may enter through storm drains into coastlines. The developing countries discharge huge quantities of raw sewage into the oceans, causing major pollution effects and health disorders. The improper maintenance and leakage of septic tanks contaminate groundwater and flow downhill into coastal areas. The fecal contamination in coastal water leads to dangerous pathogens, which cause harmful health effects. The cruise ship with 3000-7000 passengers is a floating city that discharges gallons of sewage, gray water, and tons of solid waste directly, without any treatment, into the ocean, causing severe effects to marine life. Industries without any proper treatment discharge their effluents and sludge directly into the marine ecosystem. These effluents have harmful toxicants and heavy metals that have a huge impact on marine biota. Besides, a number of solid waste management landfill areas are present near the coastal areas.

## Green House Gases

$CO_2$ is an important greenhouse gas which is released into the environment due to the burning of fossil fuels such as oil, gas, and coal for producing electricity and to run machineries or vehicles. When we burn oil, gas, and coal to produce electricity and run machinery, we add massive amounts of $CO_2$ to the environment. Methane is released from livestock, paddy farming, and landfill emissions. The CFC (Chloro Fluro Carbons) used in refrigerators emits greenhouse gases. Deforestation also enhances greenhouse gases.

## Shipping

Without environmental concern, 75% of ships are dismantled and scrapping processes are carried out manually in the coastal region of India. It leads to nearly 100 tons of heavy metals and toxic chemicals, such as paints coated with metals. Shipbreaking also causes major pollution in the Bay of Bengal. Cruise ships release untreated waste into the marine ecosystem.

## Radioactive Waste

The testing of weapons releases more radioactive substances into the marine ecosystem. Moreover, the radionuclide is discharged in the Bay of Bengal from offshore power plants in the form of liquid discharge and atmospheric routes.

## Noise

Ship traffics, oil exploration activities, acoustic research, military, air guns, fish finders, doppler current profilers, sonar, and high-resolution seafloor mapping devices create noise at a higher frequency in the marine environment.

## IMPACTS OF POLLUTANTS ON MARINE ECOSYSTEM AND BIODIVERSITY

## Oil Spills

Oil spill floats on the water surface and coat the fur and feathers of marine birds that reduces their insulation, flying ability, and make them sensitive to temperature fluctuations. Oil spills reduce the feeding rate, cause damage the fish eggs and larvae, lead to the failure of circulation, loss of hatching ability, starvation, and increased mortality. The marine mammals are severely affected by oil spills that cause metabolic disorders and organ failure. Due to the impact of oil spills, dolphins are easily prone to a pathogenic bacterial infection that could possibly lead to death [8 - 15]. The animals may ingest the oil that affects the digestive tract, kidneys, and liver, which leads to the death of organisms by the failure of organs and dehydration. The crude or refined oil spill has toxicant hydrocarbons that cause undesirable health effects in marine organisms, such as impairment of feeding and decrease in immunity, which increases their susceptibility to a disease. It may be prone to endangering conditions of animals in the affected area. It also reduces light penetration, which limits the photosynthesis of the marine plants and phytoplanktons. It also affects recreational uses such as bathing and swimming.

## Marine Debris

The marine ecosystem is severely affected by the debris that includes plastic, metal, glass, and paper. Most of the sea birds and mammals are affected by entanglement or ingestion of marine debris. The impacts of entanglement include suffocation, drowning, decreased ability to catch food and avoiding predators [16 - 19]. Their ingestion causes destruction of the gut, and organ failure due to toxicity caused by the toxicants that are absorbed. The ingestion or entrapment of marine debris causes the death of the animals. Many marine animals are already threatened or endangered due to the presence of marine debris, which is a direct hazard. Beaches are the important interface between land and sea, which is polluted with marine debris when waste is transported into the ocean by runoff and wind. Tourism and economy are also affected by marine debris, which causes undesirable changes in the region [20 - 22].

## Heavy Metals

Heavy metals accumulation in marine organisms leads to biomagnification problems at all trophic levels. For example, mercury accumulation leads to Minamata disease (numbness of limbs, visual/hearing/speech disorder) while that of cadmium leads to Itai-Itai disease (Kidney dysfunction and osteomalacia). The consumption of heavy metal contaminated seafood by pregnant mothers leads to a cognitive deficit in new born children. The prenatal exposure to mercury may cause Autistic Spectrum Disease in children. The lead exposure may cause a loss of vision and ocular plumbism in children [23 - 26]. The heavy metals cadmium, mercury, arsenic, and lead cause spontaneous abortion, premature birth, and early neonatal death in pregnant women. Heavy metal poisoning leads to loss of intelligence in children. The trace metals contamination is possible in beach soil, and groundwater during tide and floods [27 - 31].

## Nutrient Enrichment

The enrichment of phosphorus and nitrogen causes the eutrophication problem.. Thisresults in algal bloom, which releases toxins. Algal blooms prevent sunlight, clouding the water and smothering the coral. Neurotoxins that are produced by some algal blooms may cause death in marine animals and human beings. They also causes poor aesthetics, odor, and reduced oxygen level, which decline the fisheries. The water is not used for recreation, fishing and swimming purposes and causes a loss in tourism and seafood revenues.

## Plastic Debris

Marine wildlife is affected by plastic debris through ingestion, entanglement, and bioaccumulation. The plastic litter was ingested by 250 marine species, which cause physical damage or blockage in the intestinal tract, reduction of food intake, organ failures, and cause death [32, 33]. The impacts of plastic on marine organisms were observed for different biota, including seabirds, fish, mammals, sea turtles, and a range of invertebrates [34 - 43]. Entanglement threatens the survival and persistence of some species. Mostly the younger and immature animals are entangled in nets, whereas the adults are entangled in line [44 - 47].

Ghost nets also entangle turtles, sharks, crocodiles, crabs, lobsters, and numerous other species. The polychlorinated biphenyl (PCB) causes reproductive disabilities. The ingestion of plastics in seabirds may cause energy effects during flying [48 - 52]. Microplastics are of great concern that moves up through the food chain. The abundance of plastic fragments, which are less than 5 mm in diameter or small virgin plastic pellets, pose a substantial threat to marine biota [53 - 55]. Plastic pollution reduces recreational activities, such as swimming and water sports. It loses its aesthetic value and revenue due to contamination.

## Toxic Chemicals

The toxicants in the marine ecosystem cause either acute or chronic effects in the exposed organisms. They cause impairment in growth, development, reproduction, and finally death. Some toxicants bioaccumulated in marine organisms and through the food chain, affect the other trophic levels too. Many pesticides, plastics, and detergents may contain endocrine disrupting chemicals that disturb the normal activity of the endocrine system. They disturb the development and reproductive activities of organisms. The toxic chemicals cause ulcers, fin erosion, tumors, and diseases in fish. Radioactive waste causes genetic disorders, which severely affects the marine biota.

## Sewage And Industrial Discharge

Raw sewage, which includes toilets and other household wastewater, contains toxic chemicals, pathogenic microbes, and nutrients. The industries discharge the untreated effluents with more toxicants into the coastal region that has severe effects on marine biodiversity. It affects activities such as boating and swimming in the local area. It may induce gastrointestinal diseases to the beachgoers and swimmers. Fishes are highly affected by pathogens, which cause health disorders to human beings, and moreover, the economy is reduced.

## Landfills

The presence of landfills to nearby seashore may contaminate soil, and groundwater by leachate production. It results in the biomagnification problem through food chains by seafoods, vegetables, and plants cultivated in coastal areas using contaminated water.

## Greenhouse Gases

The excess $CO_2$ dissolved in water form carbonic acid that lowers the pH level and causes ocean acidification that alters the natural ecosystem balance. The marine animals could not build shells due to acidity, which poses a severe threat to shellfish industries. It causes climate change effects, which increase temperature globally and results in the melting of glaciers, increasing sea level, and flooding, which is a threat to oceanic wildlife. It may cause severe impact on mangrove swamps and salt marshes. Some sensitive species, such as coral, are not able to survive if there is a change in ocean temperature and depth. It may cause coral bleaching and affect the survival rate of different types of coral reefs globally. The changes in the marine ecosystem threatens life of marine animals. It also impacts the migration and breeding of many marine organisms, such as fishes and birds.

## Noise

It causes adverse effects on marine animals such as stress, hearing loss, impaired communication needs for reproduction and navigation, disturbed ability to detect prey and avoid predators. Sonar use may affect marine mammals and cause traumatic injuries and death. The low frequency of noise coming out from ships may disturb the communication skills of animals.

## CONTROL MEASURES FOR MARINE POLLUTION

### Oil Spills

We can reduce the utilization of petroleum, which prevents marine pollution by oily substances. Use of alternative biofuels, such as solar, wind, bioenergy *etc.* help to control oil pollutants. Crude or refined oil, car wash, and other household products based on oil should not be discharged in storm strains. Regular maintenance of engines in motorboats may reduce the release of oil into marine water. The oil tankers used for transportation should be fitted with a double hull that may reduce spillage during a shipwreck. Efficient eco-friendly biological treatment, adsorbents, and gelling agents should be needed to clean up an oil spill. The beaches contaminated with oil must be washed with high pressurized water. The animals exposed to oil pollution should be treated properly to recover.

## Marine Debris

To limit trash, the three Rs should be followed: reduce, reuse, and recycle. All the trash must be thrown in a closed bin instead of littering. The litter around homes or public places should be collected. Waste should be either recycled or reused, or managed by eco-friendly waste management methods.

## Nutrients

Source reduction from the agricultural fields, industrial discharge, and household waste must be carried out to avoid entry of nutrients into the marine ecosystem by runoff. Using advanced agricultural practices and buffer strip around the field by grass may reduce the runoff. The pollutants must always be kept out of storm drains. Dumping of household waste into a storm drain must be avoided. Car wash should be recycled before disposal. Permeable surfaces must be used to allow rainwater to seep into the ground and filtering out the contaminants. Plants are irrigated by redirecting stormwater. Always purchase low nutrient detergents or use ecofriendly products. National policy should be developed to reduce nutrient enrichment in the marine ecosystem. Regular monitoring of nutrient levels and assessment surveys should be done in the marine region to reduce nutrients.

## Plastics

We can use cloth or jute bags for purchase, reduce the packaging for products, and use a refillable water bottle during shopping in order to reduce plastic waste. The utilization of eco-friendly biodegradable/photodegradable plastics is useful to solve the problem caused by plastics. Proper biodegradation of plastics should be done by efficient microbes. Recycling and reuse of plastics also may solve the problem caused by plastics. Dumping of plastics in storm drains should be avoided. Strict guidelines should be given to beachgoers to avoid the use and dumping of plastics in the coastal region.

## Toxic Chemicals

The toxic chemicals should be removed or bioremediated by the source itself in order to avoid contamination into the marine ecosystem. Dredging up the contaminated area should be done to prevent the impacts of toxicants on the environment. Advanced eco-safe technologies must be used to limit the entry of toxicants into the ocean. Radioactive waste should be dumped in a buffer zone. Heavy metals should be recovered by advanced technical methods before discharging the waste into the marine system.

## Sewage and Industrial Effluent

The effluents and sewage should be treated properly before disposal. The industries could reuse the treated wastewater for either processing or irrigation. Regular monitoring and system maintenance of treatment plants and household septic tanks may prevent accidental releases. Cruise ships should treat their waste before disposal into the marine ecosystem. Standard limits should be followed before the discharge of wastewater into the coastal area. Rapid diagnostic tools are used to track sewage contamination and immediate remedial measures are followed to control marine pollution.

## Green House Gases

We can utilize alternative energy sources, such as solar and wind energy, to reduce greenhouse gas emissions and find out alternatives to CFC used in refrigerators. Air pollution from coal power plants could be reduced by installing scrubbers.

## Kyoto Protocol

It is a worldwide treaty that extends the 1992 United Nations Framework Convention on Climate Change (UNFCCC), which cautioned decreasing global warming by the discount of greenhouse gasoline emissions inside the surroundings. The Kyoto Protocol applies to the six greenhouse gases listed in Annex A: Carbon dioxide ($CO_2$), Methane ($CH_4$), Nitrous oxide ($N_2O$), Hydrofluorocarbons (HFCs), Perfluorocarbons (PFCs), and Sulphur hexafluoride ($SF_6$) [56].

## Noise

The sensitive marine habitats should be protected by implementing noise mitigation efforts and limiting exploration in biologically important zones. The regular maintenance of boat motors could help reduce noise output.

The following Table **1** represents the important objectives of international conventions related to the control of marine pollution [57 - 74].

**Table 1. International Conventions and their main objectives.**

| International Conventions | Objectives |
|---|---|
| OILPOL, 1954 | It was the first global treaty for the prevention of pollution of the sea by oil. It prohibited the intentional discharge of oil and oily combos from vessels in distinct ocean regions. |
| Geneva Conventions, 1958 | It has the objectives of environmental protection of the ocean from oil pollution *via* oil pipelines or continental shelf development. |
| London Convention, 1972 | It is an International Convention for the prevention of marine pollution *via* the dumping of wastes or other counts. Dumping of some wastes calls for a prior special or trendy allow. |
| Stockholm declaration, 1972 | The United Nations Conference on the Human Environment (UNCHE), held in Stockholm in June 1972, has special sections on marine pollutants. |
| MARPOL, 1973 | It is the international convention for the prevention of pollution from ships. Torrey Canyon disaster accelerated the formation of this conference that addresses pollutants and dumping from ships due to operational losses or accidents. |
| UNCLOS, 1982 | The United Nations Convention on the Law of the Sea turned into followed on 10 December 1982, and got here into pressure on 16 November 1994. It makes a specialty of marine environmental safety from sea-bed activities and the prevention of pollution from ships and land-based sources of pollution, as well as dumping and pollution switch from one country to some other. |
| Cartagena convention, 1986 | It turned into followed in 1983 and entered into pressure in 1986 which addresses pollution from ships, dumping at sea, and land-based resources of pollution in the wider Caribbean area. |
| Basel Convention, 1989 | It is an international treaty for the control of transboundary moves of risky wastes and their disposal among nations. |
| Agenda 21, 1992 | The Earth Summit held in Rio de Janeiro, Brazil in June 1992 was also very essential for the environmental and developmental issues. Chapter 17 of this document aimed for the safety of the oceans. It mentioned the need to prevent marine pollutants from vessels, which include unlawful discharges and pollutants due to ships in sensitive regions. |
| OSPAR, 1992 | It is the convention for the protection of the marine environment of the North East Atlantic. It was held in Paris on 22 September 1992. It addressed discharges, misplaced and discarded fisheries substances from vessels, land-primarily based wastes, and leisure littering. |
| Helsinki Convention, 1992 | It was the convention for the protection of marine surroundings of the Baltic Sea location. It addressed marine pollutants from all assets. It is an international convention that pursuits the prevention and elimination of pollution of the Baltic Sea. |
| UNFCC, 1994 | It is a UN Framework Convention on Climate Change. It aimed for the prevention of human interference with the climate system. |
| Barcelona Convention, 1995 | It focused on the land and sea waste from dumping, runoff, and discharges (including plastics) within the Mediterranean Sea area. |

(Table 1) cont.....

| International Conventions | Objectives |
|---|---|
| **POPs, 2001** | It is the Stockholm Convention on Persistent Organic Pollutants. It is an international treaty that focused on protecting the health of human beings and the surroundings from the impacts of POPs. |
| **SAICM**, 2002 | It is the Strategic Approach to International Chemicals Management. It is a world summit on sustainable development. It has a goal to make certain chemical substances that are produced and utilized in ways that reduce their harmful effects. It has a key position in selling more secure chemical substances policy *via* toxics reduction. |
| **Minamata convention on Mercury,** 2013 | It aimed for the protection of human health and the environment from anthropogenic emissions and releases of mercury and mercury-based compounds. |
| **G 7 Action Plan, 2015** | G7 (Canada, France, Germany, Italy, Japan, U.K., and U.S.) has a movement plan to combat marine litter. The countries reduce and get rid of marine litter by the implementation of national action plans. |
| **G 20 action plan, 2017** | Its goals for the action plan on marine litter promoted waste prevention, control, efficient wastewater remedy and management of stormwater and developing attention. |

The marine ecosystem has significant ecological and economic values. Most of the marine organisms are highly sensitive to marine pollution. Therefore, there is a need to protect the marine ecosystem from it. Environmental policies related to the protection of the marine ecosystem should be strictly followed. The pollution effects may be reduced by regular monitoring and environmental awareness policies. Industries should discharge their effluents after efficient wastewater treatments. The use and discharge of some dangerous pollutants have been banned. Oil spills should be treated by advanced technology. It is a must to protect the ocean's health for future generations.

## CONSENT FOR PUBLICATION

Not applicable.

## CONFLICT OF INTEREST

The authors confirm that this chapter contents have no conflict of interest.

## ACKNOWLEDGEMENTS

Declared none.

## REFERENCES

[1]　Ruiz-Villarreal M, González-Pola C, Diaz del Rio G, *et al.* Oceanographic conditions in North and

Northwest Iberia and their influence on the Prestige oil spill. Mar Pollut Bull 2006; 53(5-7): 220-38. [http://dx.doi.org/10.1016/j.marpolbul.2006.03.011] [PMID: 16698046]

[2]     Kaly UL. Review of Land-based sources of pollution to the coastal and marine environments in the BOBLME Region theme report: Bay of Bengal Large Marine Ecosystem (BOBLME) : GCP/RAS/179/ WBG.10 FAO-BOBLME Programme. 2004.

[3]     Staples D. Transboundary Diagnostic Analysis of the Bay of Bengal Large Marine Ecosystem: BOBLME, FAO 2012.https://www.boblme.org/documentRepository/BOBLME-2012-TDA-Volume_1

[4]     Ramesh R, Nammalwar P, Gowri VS. Database on Coastal Information of Tamil Nadu. Chennai: Institute for Ocean Management Anna University Chennai 2008.

[5]     Siddiquee NA, Parween S, Quddus MMA, Barua P. Heavy metal pollution in sediments at ship breaking area of Bangladesh. In: Subramanian V, Ed. Coastal Environments: Focus on Asian Regions. Dordrecht: Springer 2009; pp 78-87.

[6]     Shuval H. Estimating the global burden of thalassogenic diseases: human infectious diseases caused by wastewater pollution of the marine environment. J Water Health 2003; 1(2): 53-64. [http://dx.doi.org/10.2166/wh.2003.0007] [PMID: 15382734]

[7]     American Academy of Pediatrics. Chemical found in crude oil linked to congenital Heart disease: Fetal exposure to solvents may damage heart. Science Daily 2011. http://www.sciencedaily.com/ releases/2011/04/ 110430133127.htm

[8]     González-Doncel M, González L, Fernández-Torija C, Navas JM, Tarazona JV. Toxic effects of an oil spill on fish early life stages may not be exclusively associated to PAHs: studies with Prestige oil and medaka (Oryzias latipes). Aquat Toxicol 2008; 87(4): 280-8. [http://dx.doi.org/10.1016/j.aquatox.2008.02.013] [PMID: 18405983]

[9]     Sørhus E, Incardona JP, Karlsen Ø, et al. Crude oil exposures reveal roles for intracellular calcium cycling in haddock craniofacial and cardiac development. Sci Rep 2016; 6: 31058. [http://dx.doi.org/10.1038/srep31058] [PMID: 27506155]

[10]    Sørhus E, Incardona JP, Furmanek T, et al. Novel adverse outcome pathways revealed by chemical genetics in a developing marine fish. eLife 2017; 6: e20707. [http://dx.doi.org/10.7554/eLife.20707] [PMID: 28117666]

[11]    Fodrie FJ, Able KW, Galvez F, et al. Integrating organismal and population responses of estuarine fishes in Macondo spill research. Biosci 2014; 64(9): 778-88. [http://dx.doi.org/10.1093/biosci/biu123]

[12]    Hicken CE, Linbo TL, Baldwin DH, et al. Sublethal exposure to crude oil during embryonic development alters cardiac morphology and reduces aerobic capacity in adult fish. Proc Natl Acad Sci USA 2011; 108(17): 7086-90. [http://dx.doi.org/10.1073/pnas.1019031108] [PMID: 21482755]

[13]    Langangen Ø, Olsen E, Stige LC, et al. The effects of oil spills on marine fish: Implications of spatial variation in natural mortality. Mar Pollut Bull 2017; 119(1): 102-9. [http://dx.doi.org/10.1016/j.marpolbul.2017.03.037] [PMID: 28389076]

[14]    Van Dolah FM, Neely MG, McGeorge LE, et al. Seasonal variation in the skin transcriptome of common bottlenose dolphins (Tursiops truncatus) from the northern Gulf of Mexico. PLoS One 2015; 10(6): e0130934. [http://dx.doi.org/10.1371/journal.pone.0130934] [PMID: 26110790]

[15]    Venn-Watson S, Colegrove KM, Litz J, et al. Adrenal gland and lung lesions in Gulf of Mexico common bottlenose dolphins (Tursiops truncatus) found dead following the deepwater horizon oil spill. PLoS One 2015; 10(5): e0126538. [http://dx.doi.org/10.1371/journal.pone.0126538] [PMID: 25992681]

[16]    Thompson RC, Moore CJ, vom Saal FS, Swan SH. Plastics, the environment and human health: current consensus and future trends. Philos Trans R Soc Lond B Biol Sci 2009; 364(1526): 2153-66.

[http://dx.doi.org/10.1098/rstb.2009.0053] [PMID: 19528062]

[17]  Derraik JGB. The pollution of the marine environment by plastic debris: a review. Mar Pollut Bull 2002; 44(9): 842-52.
[http://dx.doi.org/10.1016/S0025-326X(02)00220-5] [PMID: 12405208]

[18]  Gregory MR. The hazards of persistent marine pollution: drift plastics and conservation islands. J R Soc N Z 1991; 21: 83-100.
[http://dx.doi.org/10.1080/03036758.1991.10431398]

[19]  Laist DW. Impacts of marine debris: entanglement of marine life in marine debris including a comprehensive list of species with entanglement and ingestion records. In: Coe JM, Rogers DB, Eds. Marine Debris. Springer Series on Environmental Management. New York: Springer 1997; pp 99-139.
[http://dx.doi.org/10.1007/978-1-4613-8486-1_10]

[20]  Bjorndal KA, Bolten AB, Lagueux CJ. Ingestion of marine debris by juvenile sea turtles in coastal Florida habitats. Mar Pollut Bull 1994; 28: 154-8.
[http://dx.doi.org/10.1016/0025-326X(94)90391-3]

[21]  Ryan PG. Effects of ingested plastic on seabird feeding: evidence from chickens. Mar Pollut Bull 1988; 19: 125-8.
[http://dx.doi.org/10.1016/0025-326X(88)90708-4]

[22]  Spear LB, Ainley DG, Ribic CA. Incidence of plastic in seabirds from the tropical Pacific, 1984–91: relation with distribution of species, sex, age, season, year and body weight. Mar Environ Res 1995; 40: 123-46.
[http://dx.doi.org/10.1016/0141-1136(94)00140-K]

[23]  Rahman J, Vajanapoom N, Van Der Putten M, Rahman N. Impact of Coastal Pollution on Childhood Disabilities and Adverse Pregnancy Outcomes: The Case of Bangladesh. Int J Med Public Health 2012; 2(3): 12-20.
[http://dx.doi.org/10.5530/ijmedph.2.3.4]

[24]  Grandjean P, Weihe P, White RF, *et al.* Cognitive deficit in 7-year-old children with prenatal exposure to methylmercury. Neurotoxicol Teratol 1997; 19(6): 417-28.
[http://dx.doi.org/10.1016/S0892-0362(97)00097-4] [PMID: 9392777]

[25]  Grandjean P, Murata K, Budtz-Jørgensen E, Weihe P. Cardiac autonomic activity in methylmercury neurotoxicity: 14-year follow-up of a Faroese birth cohort. J Pediatr 2004; 144(2): 169-76.
[http://dx.doi.org/10.1016/j.jpeds.2003.10.058] [PMID: 14760255]

[26]  Adams JB, Baral M, Geis E, *et al.* The severity of autism is associated with toxic metal body burden and red blood cell glutathione levels. J Toxicol 2009; 2009: 532640.
[http://dx.doi.org/10.1155/2009/532640] [PMID: 20107587]

[27]  Khan MAA, Khan YSA. Trace metal in littoral sediments from the north east coast of the bay of bengal along the ship breaking area, Chittagong, Bangladesh. J Biol Sci 2003; 3(11): 1050-7.
[http://dx.doi.org/10.3923/jbs.2003.1050.1057]

[28]  Alam MG, Snow ET, Tanaka A. Arsenic and heavy metal contamination of vegetables grown in Samta village, Bangladesh. Sci Total Environ 2003; 308(1-3): 83-96.
[http://dx.doi.org/10.1016/S0048-9697(02)00651-4] [PMID: 12738203]

[29]  Meliker JR, Wahl RL, Cameron LL, Nriagu JO. Arsenic in drinking water and cerebrovascular disease, diabetes mellitus, and kidney disease in Michigan: a standardized mortality ratio analysis. Environ Health 2007; 6: 4.
[http://dx.doi.org/10.1186/1476-069X-6-4] [PMID: 17274811]

[30]  Milton AH, Smith W, Rahman B, *et al.* Chronic arsenic exposure and adverse pregnancy outcomes in bangladesh. Epidemiology 2005; 16(1): 82-6.
[http://dx.doi.org/10.1097/01.ede.0000147105.94041.e6] [PMID: 15613949]

[31]  Trasande L, Landrigan PJ, Schechter C. Public health and economic consequences of methyl mercury

toxicity to the developing brain. Environ Health Perspect 2005; 113(5): 590-6.
[http://dx.doi.org/10.1289/ehp.7743] [PMID: 15866768]

[32]    Wright SL, Thompson RC, Galloway TS. The physical impacts of microplastics on marine organisms: a review. Environ Pollut 2013; 178: 483-92.
[http://dx.doi.org/10.1016/j.envpol.2013.02.031] [PMID: 23545014]

[33]    Müller C, Townsend K, Matschullat J. Experimental degradation of polymer shopping bags (standard and degradable plastic, and biodegradable) in the gastrointestinal fluids of sea turtles. Sci Total Environ 2012; 416: 464-7.
[http://dx.doi.org/10.1016/j.scitotenv.2011.10.069] [PMID: 22209368]

[34]    van Franeker JA, Blaize C, Danielsen J, *et al.* Monitoring plastic ingestion by the northern fulmar *Fulmarus glacialis* in the North Sea. Environ Pollut 2011; 159(10): 2609-15.
[http://dx.doi.org/10.1016/j.envpol.2011.06.008] [PMID: 21737191]

[35]    Ramos J, Barletta M, Costa M. Ingestion of nylon threads by Gerreidae while using a tropical estuary as foraging grounds. Aquat Biol 2012; 17: 29-34.
[http://dx.doi.org/10.3354/ab00461]

[36]    Dantas D, Barletta M, Ramos J, Lima A, Costa M. Seasonal diet shifts and overlap between two sympatric catfishes in an estuarine nursery. Estuaries Coasts 2013; 36: 237-56.
[http://dx.doi.org/10.1007/s12237-012-9563-2]

[37]    Choy CA, Drazen JC. Plastic for dinner? Observations of frequent debris ingestion by pelagic predatory fishes from the central North Pacific. Mar Ecol Prog Ser 2013; 485: 155-63.
[http://dx.doi.org/10.3354/meps10342]

[38]    Laist DW. Overview of the biological effects of lost and discarded plastic debris in the marine environment. Mar Pollut Bull 1987; 18: 319-26.
[http://dx.doi.org/10.1016/S0025-326X(87)80019-X]

[39]    Page B, McKenzie J, McIntosh R, *et al.* Entanglement of Australian sea lions and New Zealand fur seals in lost fishing gear and other marine debris before and after Government and industry attempts to reduce the problem. Mar Pollut Bull 2004; 49(1-2): 33-42.
[http://dx.doi.org/10.1016/j.marpolbul.2004.01.006] [PMID: 15234872]

[40]    Lazar B, Gračan R. Ingestion of marine debris by loggerhead sea turtles, *Caretta caretta*, in the Adriatic Sea. Mar Pollut Bull 2011; 62(1): 43-7.
[http://dx.doi.org/10.1016/j.marpolbul.2010.09.013] [PMID: 21036372]

[41]    Schuyler Q, Hardesty BD, Wilcox C, Townsend K. Global analysis of anthropogenic debris ingestion by sea turtles. Conserv Biol 2014; 28(1): 129-39.
[http://dx.doi.org/10.1111/cobi.12126] [PMID: 23914794]

[42]    Chiappone M, Dienes H, Swanson DW, Miller SL. Impacts of lost fishing gear on coral reef sessile invertebrates in the Florida Keys National Marine Sanctuary. Biol Conserv 2005; 121: 221-30.
[http://dx.doi.org/10.1016/j.biocon.2004.04.023]

[43]    Boland RC, Donohue MJ. Marine debris accumulation in the nearshore marine habitat of the endangered Hawaiian monk seal, *Monachus schauinslandi* 1999-2001. Mar Pollut Bull 2003; 46(11): 1385-94.
[http://dx.doi.org/10.1016/S0025-326X(03)00291-1] [PMID: 14607537]

[44]    Karamanlidis AA, Androukaki E, Adamantopoulou S, *et al.* Assessing accidental entanglement as a threat to the Mediterranean monk seal *Monachus monachus*. Endanger Species Res 2008; 5: 205-13.
[http://dx.doi.org/10.3354/esr00092]

[45]    Votier SC, Archibald K, Morgan G, Morgan L. The use of plastic debris as nesting material by a colonial seabird and associated entanglement mortality. Mar Pollut Bull 2011; 62(1): 168-72.
[http://dx.doi.org/10.1016/j.marpolbul.2010.11.009] [PMID: 21112061]

[46]    Henderson JR. A pre- and post-MARPOL Annex V summary of Hawaiian monk seal entanglements

and marine debris accumulation in the northwestern Hawaiian Islands, 1982-1998. Mar Pollut Bull 2001; 42(7): 584-9.
[http://dx.doi.org/10.1016/S0025-326X(00)00204-6] [PMID: 11488238]

[47]   Hanni KD, Pyle P. Entanglement of pinnipeds in synthetic materials at South-east Farallon Island, California, 1976–1998. Mar Pollut Bull 2000; 40: 1076-81.
[http://dx.doi.org/10.1016/S0025-326X(00)00050-3]

[48]   Gunn R, Hardesty BD, Butler J. Tackling 'ghost nets': local solutions to a global issue in northern Australia. Ecol Manage Restor 2010; 11: 88-98.
[http://dx.doi.org/10.1111/j.1442-8903.2010.00525.x]

[49]   Wilcox C, Hardesty B, Sharples R, Griffin D, Lawson T, Gunn R. Ghost net impacts on globally threatened turtles, a spatial risk analysis for northern Australia. Conserv Lett 2013; 6: 247-54.
[http://dx.doi.org/10.1111/conl.12001]

[50]   Wilcox C, Van Sebille E, Hardesty BD. Threat of plastic pollution to seabirds is global, pervasive, and increasing. Proc Natl Acad Sci USA 2015; 112(38): 11899-904.
[http://dx.doi.org/10.1073/pnas.1502108112] [PMID: 26324886]

[51]   Velzeboer I, Kwadijk CJAF, Koelmans AA. Strong sorption of PCBs to nanoplastics, microplastics, carbon nanotubes, and fullerenes. Environ Sci Technol 2014; 48(9): 4869-76.
[http://dx.doi.org/10.1021/es405721v] [PMID: 24689832]

[52]   Rochman CM, Browne MA, Halpern BS, *et al.* Policy: Classify plastic waste as hazardous. Nature 2013; 494(7436): 169-71.
[http://dx.doi.org/10.1038/494169a] [PMID: 23407523]

[53]   Barnes DKA, Galgani F, Thompson RC, Barlaz M. Accumulation and fragmentation of plastic debris in global environments. Philos Trans R Soc Lond B Biol Sci 2009; 364(1526): 1985-98.
[http://dx.doi.org/10.1098/rstb.2008.0205] [PMID: 19528051]

[54]   Goldstein MC, Rosenberg M, Cheng L. Increased oceanic microplastic debris enhances oviposition in an endemic pelagic insect. Biol Lett 2012; 8(5): 817-20.
[http://dx.doi.org/10.1098/rsbl.2012.0298] [PMID: 22573831]

[55]   Wyles, Pahl KJS and Thompson RC. Perceived risks and benefits of recreational visits to the marine environment: Integrating impacts on the environment and impacts on the visitor. Ocean Coast Manage 2014; 8: 53-63.

[56]   Kyoto Protocol to the United Nations Framework Convention on Climate Change 1998. https://unfccc.int/resource/docs/convkp/kpeng.pdf

[57]   International Convention for the Prevention of Pollution of the Sea by Oil United Nations, Treaty Series 1954; 3-327.

[58]   Convention on the High Seas. United Nations. Treaty Series 1958; 450: 11.

[59]   Convention on the prevention of marine pollution by dumping of wastes and other matter, Dec.29, 1972, 26 U.S.T. 2403, 1046 U.N.T.S. 120, 11 I.L.M. 1291.

[60]   Declaration of the United Nations Conference on the Human Environment 1972. http://www.unep.org/Documents.Multilingual/Default.asp?documentid=97&articleid=1503

[61]   International convention for the prevention of pollution from Ships, Nov. 2, 1973, 12 I.L.M.1319, as amended by Protocol, Feb. 17, 1978, 17 I.L.M. 546..

[62]   United Nations Convention on the Law of the Sea. 1982. http://www.un.org/ depts/los/convention_agreements/ texts/ unclos/unclos_e.pdf

[63]   Convention for the protection and development of the marine environment in the wider caribbean region, Mar. 24, 1983, 1506 U.N.T.S. 157, 22 I.L.M. 221.

[64]   http://www.basel.int/

[65]   http://www.un.org/esa/dsd/agenda21/res_agenda21_00.shtml

[66]   Convention for the protection of the marine environment of the north east Atlantic, Sept. 22, 1993, 2354 U.N.T.S. 67, 32 I.L.M. 1069.

[67]   Convention on the protection of the marine environment of the baltic sea area, Apr. 9, 1992,1507 U.N.T.S. 167, 13 I.L.M. 546 (entered into force Jan. 17, 2000).

[68]   https://unfccc.int/process-and-meetings/the-convention/what-is-the-united-nations-f-amework-convention-on-climate-change

[69]   Convention for the protection of the Mediterranean Sea against pollution, Feb. 16, 1976, 15 I.L.M.285, revised as convention for the protection of the marine environment and the coastal region of the Mediterranean, June 10, 1995, 1102 U.N.T.S. 27 (entered into force July 9, 2004).

[70]   http://chm.pops.int/

[71]   http://www.saicm.org/

[72]   Minamata Convention on Mercury. http://mercuryconvention.org/Convention/tabid/3426/Default.aspx

[73]   https://www.env.go.jp/water/marine_litter/07_mat13_2_%EF%BC%93-2ALD

[74]   http://www.g20.utoronto.ca/2017/2017-g20-marine-litter.html

CHAPTER 10

# Multidrug Resistance *Proteus mirabilis* in Klang River, Malaysia

**N. Suryadevara[1,*], M. F. T. Lim[1], G. Shanmugam[2] and P. Ponmurugan[3]**

[1] *School of Biosciences, Faculty of Medicine, Bioscience and Nursing, MAHSA University, Malaysia*

[2] *Department of Biotechnology, Yeungnam University, Gyeongsangbuk-do, South Korea*

[3] *Biomedical Research Lab, Department of Botany, Bharathiar University, Coimbatore – 641 046, Tamil Nadu, India*

**Abstract:** Multidrug-resistance (MDR) bacteria have emerged as a public health threat in the modern era. Extended spectrum β-lactamase has emerged as the most successful resistance mechanism among the *Enterobacteriaceae* family. ESBLs are often mediated by *bla*TEM and *bla*CTX-M genes. The study evaluated the molecular characteristics of ESBL-producing *P. mirabilis* isolates from the Klang River in order to find the therapeutic options for ESBL infections. There appears to be a high prevalence of *CTX-M* and *TEM* genes among *P. mirabilis* strains in the Klang River. Therefore, rapid identification should be conducted for proper infection control and antibiotic usage. Molecular techniques serve as a useful tool in the understanding of MDR bacteria resistance.

**Keywords:** Bla$_{TEM}$/bla$_{CTX-M}$, ESBL, *K. pneumoniae*, MDR.

## INTRODUCTION

Bacteria were known to cause morbidities and mortality in the human population way before the discovery of antimicrobial agents. The invention of antibiotics dates back to the 15$^{th}$ century, when microbial infections were treated and documented well in ancient Egypt, Greece, and China [1]. The modern era of antibiotics started with Sir Alexander Fleming, who discovered penicillin in the year 1928. Since then, antibiotics have been considered to transform modern medicine and save millions of lives. However, shortly after, penicillin resistance took place, and from then onwards, each newly introduced antibiotics class has been followed by a global wave of emergent resistance, mainly originating in Europe and North America [2]. The wave further spread from the United King-

---

[*] **Corresponding author N. Suryadevara:** School of Biosciences, Faculty of Medicine, Bioscience and Nursing, MAHSA University, Malaysia; Tel: +601123048554; E-mail: nagaraja@mahsa.edu.my

**J. Senthil Kumar, P. Ponmurugan & A. Vinothkanna (Eds.)**

dom and North America across Europe and then reached Asia over more than a decade. As a result, in 2018, bacterial infections again became a threat. In order to understand the problem of antibiotic resistance, one has to first acknowledge the fact that antibiotic resistance is ancient and is a result of many organism-environment interactions. Antimicrobial resistance genes have been identified in bacterial DNA frozen in the Arctic permafrost for about 30,000 years and in bacteria in a subterranean cave isolated from the surface for more than 4 million years [2]. Most antimicrobial compounds are naturally- produced molecules. Therefore, in order to survive, co-resident bacteria exposed to these bioactive molecules must evolve self-protective resistance mechanisms as these bacteria are said to be "intrinsically" resistant to antimicrobials [3]. However, it is clear that those resistance mechanisms have been mobilized horizontally through microbial populations that led to a more problematic development of "acquired resistance" in bacteria, which mainly resulted from mutations in chromosomal genes or acquisition of external genetic determinant from the environmentally present and intrinsically resistant organisms. Both intrinsic or acquired resistance contribute to antibiotic resistance in bacteria.

The spread of multidrug-resistant (MDR) bacteria has been a global public health problem nowadays, including in South-East Asian countries like Malaysia, Singapore, Indonesia, and Thailand [4]. Infections caused by MDR bacteria are associated with increased mortality and are clinically, socially, and economically significant. For example, a recent study in Thailand showed that 43% of deaths caused by hospital-acquired MDR bacterial infections in 2010 constituted the excess mortality rate due to MDR. The situation is worsened by the incompetence of a robust antibiotic pipeline, resulting in almost untreatable infections and leaving clinicians with no reliable alternatives to treat infected patients.

Over the last few years, the presence of antibiotics as well MDR has been shown outside the clinical environment, including water, soil, and, most notably, food-producing animals. In recent years, the presence of most extended-spectrum β-lactamases (ESBLs) has been confirmed in virtually all ecological niches. Aquatic ecosystems are often recognized as reservoirs for these ESBL bacteria where these environments are highly enriched by antibiotics, specifically β-lactam antibiotics that are found in sewage, agricultural runoff, and hospital wastewater [5]. Klang Valley is the most urban region of Kuala Lumpur as it consists of relatively high population densities of cities and towns with numerous aquatic environments, such as rivers and lakes, that are most likely to be exposed to chemical and biological sources contaminated with ESBL-producing bacteria, such as *E. coli, Klebsiella pneuominae, Proteus mirabilis,* and others.

*P. mirabilis* is a Gram-negative bacterium that belongs to the *Enterobacteriaceae* family, which could be found in polluted water and soil with the characteristic features of swarming and production of urease [6]. ESBL-producing *P. mirabilis* strains emerged in the 1990s as a nosocomial infection, conferring to clonal spread and causing outbreaks. There has been sufficient evidence to support the increased morbidity and mortality with ESBL-producing *P. mirabilis* outbreaks, particularly in the nosocomial environment [7]. ESBL-producing *P. mirabilis* could lead to minor diseases like urinary tract infections (UTIs) to severe conditions such as pneumonia and bacteremia [8].

The continuous exposure of bacteria to β-lactam antibiotics induces a dynamic production and mutation of β-lactamases in the bacteria that expand their activity against newly developed β-lactam antibiotics through the production of enzymes known as ESBLs [9]. The genes responsible for producing ESBLs are typically the *bla* genes with different variants, such as TEM, OXA, SHV, CTX-M, and others. As the production of ESBLs in Asia is becoming more alarming, a new era that practices molecular characterization has been developed for better epidemiological control and reliable investigation of ESBL-producing bacteria [10, 11].

## Antimicrobial Agents

Antimicrobial is either a chemical or physical agent that kills or inhibits the growth of microorganisms, whereas the term "antibiotic" refers to any class of organic molecule that inhibits or kills microbes by specific interactions with bacterial targets, without any consideration of the source of a particular compound or class [12]. Thus, purely synthetic therapeutics are considered antibiotics. Some of the good examples are fluoroquinolones, sulphonamides, and trimethoprim. Antibiotics could also refer to the secondary metabolite produced by a bacterium that kills or inhibits other bacteria (Table 1). Therefore, the term "antibiotics" is used for antimicrobial agents, either naturally or synthetically produced.

Table 1. Some naturally produced antibiotics in soil and aquatic environment [13].

| | Microorganisms | Antibiotic Produced |
|---|---|---|
| Bacillus group | *Bacillus subtilis* | Bacitracin |
| | *Bacillus polymixa* | Polymixin |
| Streptomyces | *Streptomyces erytherus Steptomycesgriseus* | Erythromycin Streptomycin |
| | *Streptomyces rimosus Streptomyces orientalis* | Tetracycline Vancomycin |

## Classification and Mechanism of Action of Antimicrobials

There are several methods of classifying antibiotics but the most common one is based on their molecular structures, mode of action, and spectrum of activity [14]. Some common classes of antibiotics based on chemical or molecular structures include β-lactams, macrolides, tetracyclines, quinolones, aminoglycosides, sulphonamides, glycopeptides, and oxazolidinones. The spectrum of activity of the antimicrobial agents could be categorized as either narrow or broad spectrum. Only certain bacteria are inhibited by narrow-spectrum antimicrobials, for example, benzyl penicillin is active against many Gram- negative and Gram-positive cocci but has little activity against Gram-negative bacilli. On the other hand, broad-spectrum antimicrobials, for example, cephalosporins, tetracyclines, and chloramphenicol, are active against a wide range of Gram-positive and Gram-negative bacteria [14].

The antimicrobial potency of most classes of antibiotics depends on the unique feature of the bacterial structure or their metabolic processes. The most common targets of antibiotics are shown in Fig. (**1**). The mechanisms of antibiotic actions are such as [15]:

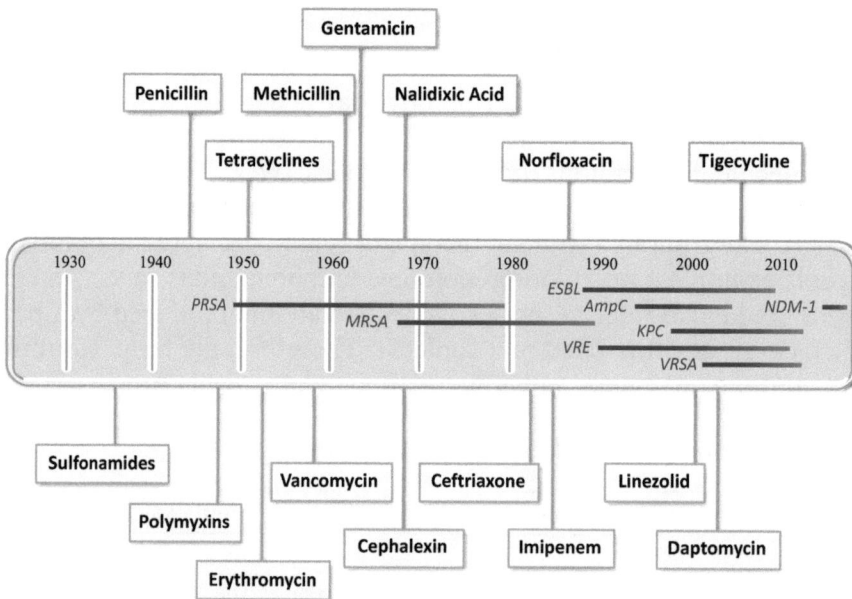

**Fig. (1).** Timeline of antimicrobial discoveries and the appearance of drug resistance (Abbreviations: AmpC, AmpC-producing EnterobacteriaceaeL ESBL, extended-spectrum β-lactamase-Producing Enterobacteriaceae; KPC, *Klebsiella pneumoniae* carbapenamase-producing Enterobacteriaceae:, MRSA, methicillin-resistant *Staphylococcus aureus*: NDM-1, New Delhi mtallo- β-lactamase-1-producing Enterobacteriaceae; PRSA, penicillin=resistant *Staphylococcus aureus*, VRE, vancomycin resistant *Enterococcus*: VRSA, vancomycin-resistant *Staphylococcus aureus*) [2].

1. Inhibition of cell wall synthesis
2. Breakdown of cell membrane structure or function
3. Inhibition of the structure and function of nucleic acids
4. Inhibition of protein synthesis

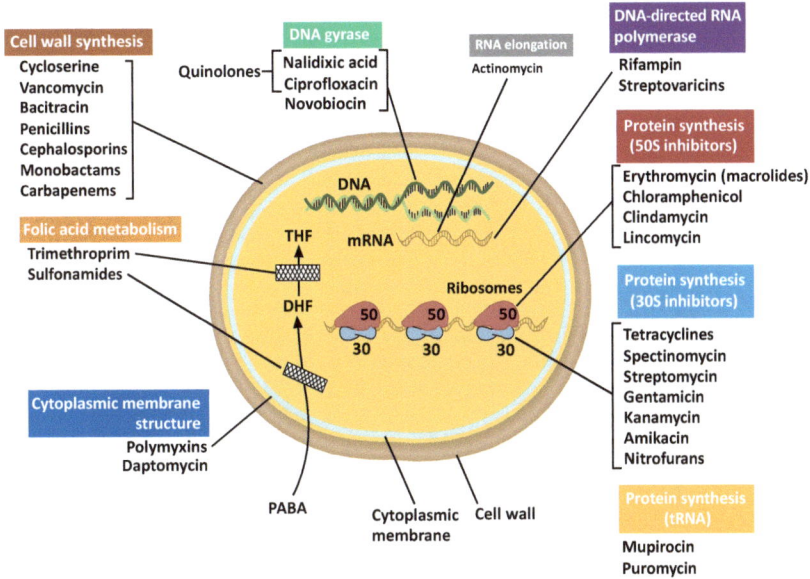

**Fig. (2).** Antibiotic target sites (Etebu and Arikekpar, 2016).

## Mechanism of Antibiotics Resistance

Bacteria have a remarkable genetic plasticity that allows them to defend themselves against a wide array of environmental threats such as antibiotic molecules. From an evolutionary perspective, bacteria develop resistance in two ways [16]:

**Intrinsic resistance** refers to the existence of genes in bacterial genomes that could generate a resistance phenotype

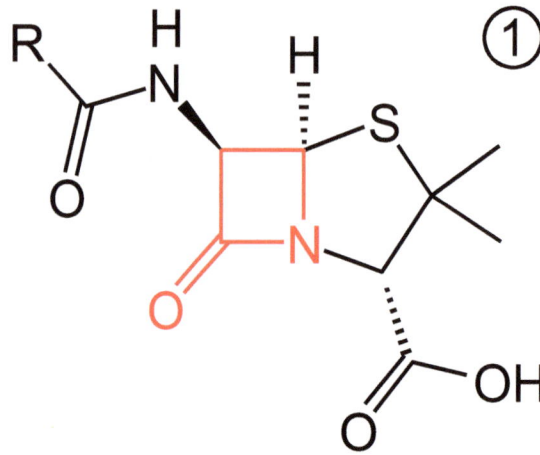

**Fig. (3).** Chemical structure of penicillins (Etebu and Arikekpar, 2016).

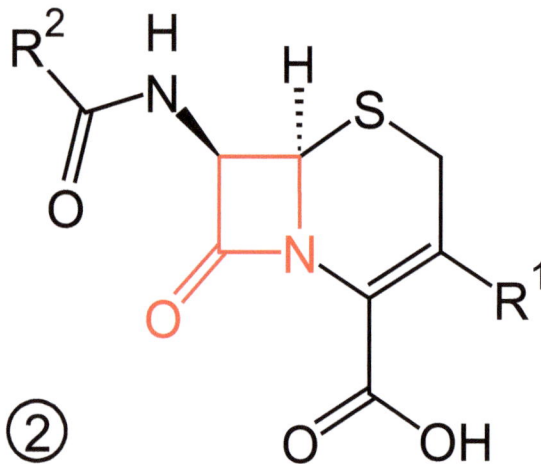

**Fig. (4).** Chemical structure of cephalosporins (Etebu and Arikekpar, 2016).

**Acquired resistance** refers to a type of resistance in which a naturally susceptible bacteria acquires mechanisms, so that it is rendered harmless by the antimicrobial agent. Mechanisms of acquired resistance include the presence of an enzyme that inactivates the antimicrobial agent, reduced uptake of antimicrobial agent, active efflux of the antimicrobial agent, and post-transcriptional or post-translational modification of the antimicrobial agent's target. Both of the mechanisms of actions and resistances of the commonly used antimicrobial agents are shown in Table **2**.

**Table 2. Modes of action and resistance mechanisms of commonly used antibiotics (Davies and Davies, 2010).**

| Antibiotic Class | Example(s) | Target | Mode(s) of Resistance |
|---|---|---|---|
| β-lactams | Penicillins (ampicillin), cephalosporins (cephamycin), penems (meropenem), monobactams (aztreonam) | Peptidoglycan biosynthesis | Hydrolysis, efflux, altered target |
| Aminoglycosides | Gentamicin, streptomycin, spectinomycin | Translation | Phosphorylation, acetylation, nucleotidylation, efflux, altered target |
| Glycopeptides | Vancomycin, teicoplanin | Peptidoglycan biosynthesis | Reprogramming peptidoglycan biosynthesis |
| Tetracyclines | Minocycline, tigecycline | Translation | Monooxygenation, efflux, altered target |
| Macrolides | Erythromycin, azithromycin | Translation | Hydrolysis, glycosylation, phosphorylation, efflux, altered target |
| Lincosamides | Clindamycin | Translation | Nucleotidylation, efflux, altered target |
| Streptogramins | Synercid | Translation | C-O lyase (type B streptogramins), acetylation (type A streptogramins), efflux, altered target |
| Oxazolidinones | Linezolid | Translation | Efflux, altered target |
| Phenicols | Chloramphenicol | Translation | Acetylation, efflux, altered target |
| Quinolones | Ciprofloxacin | DNA replication | Acetylation, efflux, altered target |
| Pyrimidines | Trimethoprim | C1 metabolism | Efflux, altered target |
| Sulfonamides | Sulfamethoxazole | C1 metabolism | Efflux, altered target |
| Rifamycins | Rifampin | Transcription | ADP-ribosylation, efflux, altered target |

## *Mechanism of Action of β-lactam Antibiotics*

Members from this class of antibiotics contain a 3-carbon and 1-nitrogen β-lactam ring that is highly reactive, as shown in Figs. (**2** and **3**).

β-lactam interferes with the proteins that are important for the synthesis of the bacterial cell wall, which, in the process, either kills or inhibits their growth. More precisely, members of β-lactam antibiotics could bind themselves to a bacterial enzyme named penicillin-binding protein (PBP) that functions in cross-linking peptide units during the synthesis of peptidoglycan (Etebu and Arikekpar, 2016).

As a result of binding to these PBP enzymes, β-lactam interferes with the synthesis of peptidoglycan, resulting in lysis and cell death. The members of the β-lactam class include penicillins, cephalosporins, monobactams, and carbapenems, as shown in Fig. (**4**). The cell wall inhibition action and spectrum of activity of β-lactam are shown in Table **3**.

**Table 3. Penicillin, cephalosporins, and carbapenems as examples of cell wall-inhibiting antimicrobials (Kayser, 2005 and Levinson, 2008).**

| Antimicrobial Group | Examples | Spectrum of Action |
|---|---|---|
| **Penicillin** | | |
| 1. Natural Penicillins | Penicillin G (injection) Penicillin V (oral) | Active mainly against Gram-positive bacteria |
| 2. Penicillinase -resistant penicillins | Cloxacillin, dicloxacillin, methicillin, nafcillin, oxacillin | Anti-Staphylococcal action |
| 3. Aminopenicillins | Amoxicillin, ampicillin, amoxicillin/clavulanic acid, ampicillin/sulbactam, bacampicillin | Active against Gram-positive and Gram-negative bacteria |
| 4. Carboxypenicillins | Carbenicillin, ticarcillin, ticarcillin/clavulanic acid | Greater activity against Gram-negative organisms. |
| 5. Ureidopenicillins and piperazinepenicillins | Mezlocillin, piperacillin, piperacillin/tazobactam | Broadest spectrum of all penicillin's |
| **Cephalosporin** | | |
| 1. 1st generation | Cefadoxil, cefazolin, cephalexin, cephalothin, cephradine | Effective against Gram-positive bacteria and few Gram-negative bacteria such as *E. coli, Klebsiella pneumoniae,* and *Proteus mirabilis* |
| 2. 2nd generation | Cefaclor, cefamandole, cefmetazole, cefoxitin, cefoprzil, cefuroxime, cefpodoxime | More effective against Gram-negative bacteria such as *Enterobacter, Klebsiella,* and *Proteus* spp. |
| 3. 3rd generation | Cefixime, cefoprazone, cefotaxime, ceftazidime, ceftriaxone | Better effect against *E. coli, Klebsiella, Acineobacter, Serratia, Enterobacter, Proteus, Providencia, Morganella,* and *Neisseria* spp. |
| 4. 4th generation | Cefepime | Better action against both Gram-positive and Gram-negative organisms. |
| **Carbapenems** | | |
| Carbapenems | Imipenem, Meropenem, Doripenem, | Broad spectrum of activity against both Gram-positive and Gram-negative microorganisms. |

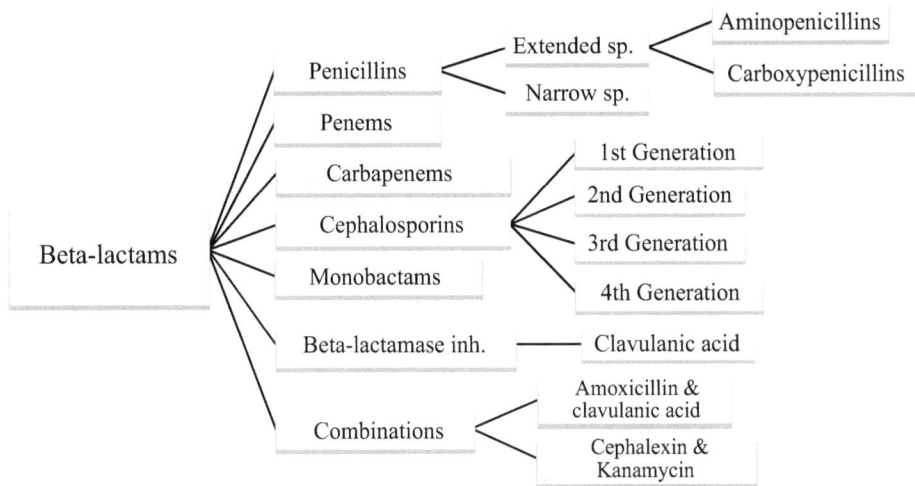

**Fig. (5).** β-lactam family of antibiotics (Dowling *et al.*, 2017).

## *Penicillin*

Some members of this family, such as ampicillin, carbenicillin, and amoxicillin, have been developed semi-synthetically with different side chains. These side chains provide the antibiotics the ability to escape the degradative capacity of certain enzymes produced by certain bacterial strains and, at the same time, help in the movement of antibiotics across the bacterial cell walls. This double ability has improved their spectrum of activity against Gram-negative bacteria. Furthermore, some penicillins, such as augmentin, which is a drug comprising of amoxicillin (antibiotic) and clavulanic acid (non-antibiotic), are able to inhibit the activity of bacterial penicillinase enzyme. Clavulanic acid functions in inhibiting the β- lactamase enzyme, therefore extending the period of antibacterial activity of the amoxicillin component even amongst penicillinase producing bacteria.

## *Cephalosporin*

Cephalosporin members are similar to penicillin in terms of their structure and their mode of action. They are subdivided into generations ($1^{st} – 4^{th}$) according to their target organisms, but the later generations are said to be more effective against Gram-negative pathogens. Cephalosporin act through their variety of side chains, which enable them to attached to different penicllin-binding proteins (PBPs), to evade the blood brain barrier, to resist breakdown by the penicillinase-producing bacterial strains, and ionize to gain entry into Gram-negative bacterial cells [15].

## Monobactam

Monobactam are a part of beta-lactam compounds, but unlike others, the β-lactam ring of monobactam stands alone (Fig. **5**). The only commercially available monobactam antibiotic is aztreonam, with a narrow spectrum of activity that is only active against aerobic Gram-negative bacteria, such as *Neisseria* and *Pseudomonas,* but not active against Gram-positive bacteria or anaerobes [15].

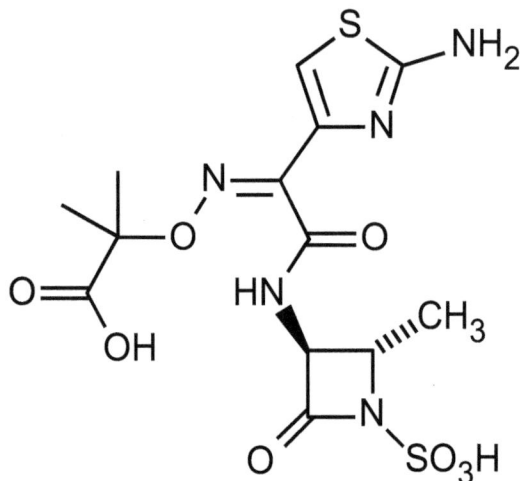

**Fig. (6).** Chemical structure of monobactams [15].

## Carbapenem

Carbapenems are able to resist the hydrolytic action of the β-lactamase enzyme. They possess the broadest spectrum and the greatest potency against Gram-positive and Gram-negative bacteria. Some of the examples are imipenem, meropenem, and ertapenem. Carbapenems are often used as the last resort antibiotics, however, the emergence of bacterial pathogens resistant to this class of antibiotics has been reported.

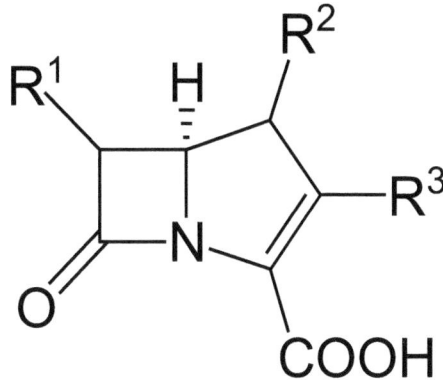

**Fig. (7).** Chemical structure of carbapenems (Etebuamd Arikekpar, 2016).

## *Mechanism of β-Lactam Resistance and ESBL*

β-lactam antibiotic resistance is facilitated by β-lactamase, that is able to hydrolyze the β-lactam ring structure of antibiotic and renders it inactive; whereas ESBLs are a branch of β-lactamase that are able to target a broader spectrum of antibiotics. They are typically plasmid-mediated enzymes that are common among *Enterobacteriaceae* that hydrolyze penicillins, third and fourth generation cephalosporins and aztreonam but not cephamycins, such as cefoxitin. They are inhibited by β-lactamase inhibitors (clavulanic acid, sulbactam, and tazobactam) [17]. β-lactamase are coded by different types of *bla* genes, where each of them differs in the amino-acid structure of the protein molecules [12]. The genes that mediate ESBLs are mainly blaSHV, blaTEM, and blaCTX-M genes. It is known that *bla* genes encoding antibiotic resistance may be placed on transferable elements such as plasmids or transposons [18]. Therefore, blaSHV and blaCTX-M are usually found encoded on the plasmid, whereas the blaTEM gene encoded on transposon can be acquired in bacteria due to the horizontal gene transfer that facilitates ESBL production [12].

## MOLECULAR CLASSIFICATION OF B-LACTAMASES

### TEM β-lactamases

TEM β-lactamases were the first ESBLs to be discovered. The variants such as TEM-1 and TEM-2 are narrow-spectrum β-lactamases that are able to hydrolyze only penicillin and the first generation of cephalosporin but are not effective against higher generations of cephalosporins, such as ceftriaxone, ceftazidime, and cefepime [9].

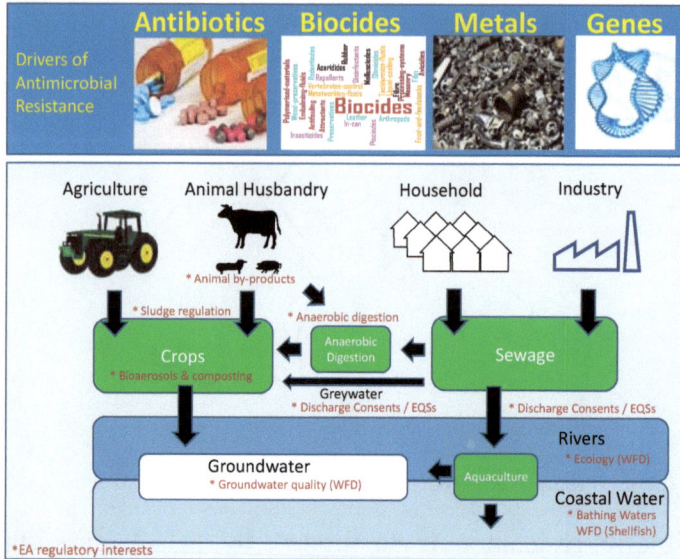

**Fig. (8).** Schematic of the hot-spots and drivers of antimicrobial resistance (Costa *et al.*, 2013).

## SHV Sulfhydryl Variants

SHVs β-lactamase variants are less common and are said to derived from *Klebsiella pneuomoniae*. Their spectrum of activity is narrower in *Enterobacteriaceae* that has resulted from the point mutation of the chromosomal SHV gene. SHV β-lactamase confers high-level resistance to ceftazidime, but not to cefotaxime and cefazolin [9].

## OXA β-lactamases

They are characterized by causing inactivity in oxacillin and related antimicrobials such as methicillin and cloxacillin. They are predominantly found in *P. aeruginosa* and many other Gram-negative bacteria [9].

## CTX-M β-lactamases

Different from the SHV- and TEM-ESBLs that are generated by amino acid substitution, CTX-M-ESBLs are acquired by the horizontal gene transfer from other bacteria, using conjugative plasmid or transposon. CTX-M ESBLs are more active against cefotaxime and ceftriaxone than ceftazidime and are better inhibited by the β-lactamase inhibitor, tazobactam, than by sulbactam and clavulanate. Five different groups of CTX-M (1, 2, 8, 9, and 25) are divided based on amino acid similarities. However, the prevalence of specific CTX-M subgroups depends on the geographic region and CTX-M-15 is currently the most widely disseminated CTX-M genotypes [9, 10].

## Antibiotic Resistance in the Environment

The environment is increasingly being recognized for its role in the global spreading of clinically relevant antibiotic resistance. An important factor in the selection and spread of resistant bacteria was the selection pressure of antibiotics present above a concentration against the microbial biocoenosis. The presence of antibiotics over a long period at subtherapeutic concentrations through numerous environmental pathways (*e.g.* water, soil, and air) (see Fig. **7**) would favor the transfer of resistance genes from environmental bacteria to bacterial pathogen that stimulates multi-drug resistance [19].

## MDR Bacteria in the Environment

MDR is defined as non-susceptibility to at least one agent in three or more antimicrobial categories [20]. The use of antimicrobial agents is claimed to drive the spread of MDR bacteria. However, the presence of MDR bacteria has been reported in populations with no apparent exposure to antimicrobials. MDR strains represent an environmental health hazard where many of them harbor virulence factors that could cause disease to the host when it is immunosuppressed or deficient in nutrition [19].

## Aquatic Environment as a Reservoir of Antimicrobial Resistance

In order to control the release of antibiotics and antibiotic-resistant bacteria in the environment, aquatic environments, be it drinking or wastewaters, are being investigated worldwide. Urban waters are frequently exposed to relatively high population densities, and therefore, are often unprotected from biological contaminants, and serve as a crucial factor in the dissemination of antibiotic resistance in the environment. The study by Tissera and Lee (2013) on the presence of ESBL species in the urban surface waters of Malaysia has shown that 19 ESBL isolates obtained were mainly *Enterobacteriaceae*. Similar conclusions can also be drawn from a study where 39 ESBL isolates were identified as *Enterobacteriaceae* from the urban river of China [21].

Moreover, 38.5% of isolates in the water sources of Iran are ESBL producers, predominantly *E. coli* [11]. Similar findings are also found in other environments, such as clinical specimens and farming environment, where most of the ESBL isolates are *Klebsiella pneumoniae* and *E. coli* [22, 23]. To track the MDR bacteria load in a water network, the water is investigated from upstream of the water body until downstream. The load of bacteria is lower in the upstream as compared to the downstream as waters are saturated with wastewater and effluents [24]. Wild strains of bacteria are mostly present on the upstream waters that should be susceptible to penicillins and cephalosporins.

## ESBL-producing *P. Mirabilis*

Since previous studies had mainly focused on common ESBL-producing bacteria such as *Klebsiella pneumoniae* and *E. coli,* there is a need to conduct a research on*P. mirabilis*. The emergence of ESBL-producing *P. mirabilis* has grown to become a concern in recent years, as they are associated with poor treatment outcome and prolonged hospitalization [25]. Over the last few years, ESBL-positive *P. mirabilis* isolates have been recovered worldwide, with a relatively high prevalence in some settings [26]. For instance, the prevalence of cefotaxime resistance among *P. mirabilis* has increased annually, from 10.1% in 1998 to 23.1% in 2003, and increased drastically in 2004, exceeding 40%. The research conducted by Endimiani *et al.* (2005) has shown ESBL positive isolates *P. mirabilis* in several bacteremia infections with good clinical efficacy against carbapenems. The options left for treatment of ESBL-associated infections are carbapenems and fourth-generation cephalosporin, such as cefepime.

## PHENOTYPIC METHODS OF ESBL

### Modified Double Disc Synergy Test (MDDST)

MDDST is the modification of the original DDST, where MDDST uses the fourth-generation cephalosporim (cefepime) that could improve the detection of ESBL in the strains which co-produce AmpC. Certain ESBL strain with co-existing AmpCβ-lactamases may give false negative results for the detection of ESBLs because the standard DDST, which uses clavulanic acid, induces a higher level of AmpC production that could lead to the resistance of third-generation cephalosporin as well their combination with clavulanic acid. The distance between the discs is critical and 20 mm center-to-center has been found to be optimal for cephalosporin 30μg discs [17].

### Whole Genomic Sequencing (WGS)

*Haemophilus influenza* was the first bacterial genome to be sequenced completely in the year 1995. Since then, more bacterial genomes have been sequenced completely or drafted. Sequencing has been improved drastically through the introduction of next-generation sequencing (NGS) in 2005, which has reduced the cost of sequencing drastically and improved the sequencing performance in small and medium-sized laboratories [27].

The steps in any bacterial genome sequencing include preparation of the sample, sequencing of DNA, sequence assembly, and bioinformatics analysis [28]. The current advances have allowed even a very little amount of sample, such as degraded original material, for DNA sequencing. By providing genotype information, an individual microbe's resistance to β-lactam drugs and resistance mechanism could be characterized effectively [29].

## Molecular Docking

After retrieving the genomic sequence, the mRNA sequence could be obtained from the Transcription and Translation Tool. The mRNA sequence is further used to generate the 3D structure of mutated proteins, such as mutated β-lactamase, for molecular docking [29]. Broadly used in modern drug design, molecular docking is a modeling technique that is used to study the protein-ligand interactions in order to identify the ligand conformation within the binding sites of a protein and to predict the affinity between the ligand and the protein by a specific scoring function [30]. The docking study paves a path for the design of novel antibiotics to cope up with the emerging MDR bacteria.

## CONSENT FOR PUBLICATION

Not applicable.

## CONFLICT OF INTEREST

The authors confirm that this chapter contents have no conflict of interest.

## ACKNOWLEDGEMENTS

Declared none.

## REFERENCES

[1]     Ventola CL. The antibiotic resistance crisis: Part 1: causes and threats. P&T 2015; 40(4): 277-83.
       [PMID: 25859123]

[2]     Molton JS, Tambyah PA, Ang BSP, Ling ML, Fisher DA. The global spread of healthcare-associated
       multidrug-resistant bacteria: a perspective from Asia. Clin Infect Dis 2013; 56(9): 1310-8.
       [http://dx.doi.org/10.1093/cid/cit020] [PMID: 23334810]

[3]     Wright GD, Poinar H. Antibiotic resistance is ancient: Implications for drug discovery. Trends
       Microbiol 2012; 20(4): 157-9.
       [http://dx.doi.org/10.1016/j.tim.2012.01.002] [PMID: 22284896]

[4]     Zellweger RM, Carrique-Mas J, Limmathurotsakul D, Day NPJ, Thwaites GE, Baker S. Southeast
       Asia antimicrobial resistance network. A current perspective on antimicrobial resistance in Southeast
       Asia. J Antimicrob Chemother 2017; 72(11): 2963-72.
       [http://dx.doi.org/10.1093/jac/dkx260] [PMID: 28961709]

[5]     Adesoji AT, Ogunjobi AA. Occurrence of multidrug-resistant bacteria in selected water distribution
       systems in Oyo State, Nigeria. Glob Vet 2013; 11(2): 214-24.

[6]     Wang JT, Chen PC, Chang SC, *et al.* Antimicrobial susceptibilities of *Proteus mirabilis*: A
       longitudinal nationwide study from the Taiwan surveillance of antimicrobial resistance (TSAR)
       program. BMC Infect Dis 2014; 14: 486.
       [http://dx.doi.org/10.1186/1471-2334-14-486] [PMID: 25192738]

[7] Luzzaro F, Perilli M, Amicosante G, *et al.* Properties of multidrug-resistant, ESBL-producing *Proteus mirabilis* isolates and possible role of β-lactam/β-lactamase inhibitor combinations. Int J Antimicrob Agents 2001; 17(2): 131-5.
[http://dx.doi.org/10.1016/S0924-8579(00)00325-3] [PMID: 11165117]

[8] Nakano R, Nakano A, Abe M, Inoue M, Okamoto R. Regional outbreak of CTX-M-2 β-lactamase-producing *Proteus mirabilis* in Japan. J Med Microbiol 2012; 61(Pt 12): 1727-35.
[http://dx.doi.org/10.1099/jmm.0.049726-0] [PMID: 22935848]

[9] Shaikh S, Fatima J, Shakil S, Rizvi SM, Kamal MA. Antibiotic resistance and extended spectrum beta-lactamases: Types, epidemiology and treatment. Saudi J Biol Sci 2015; 22(1): 90-101.
[http://dx.doi.org/10.1016/j.sjbs.2014.08.002] [PMID: 25561890]

[10] Sharma M, Pathak S, Srivastava P. Prevalence and antibiogram of extended spectrum β-lactamase (ESBL) producing gram negative bacilli and further molecular characterization of esbl producing *Escherichia coli* and Klebsiella spp. J Clin Diagn Res 2013; 7(10): 2173-7.
[http://dx.doi.org/10.7860/JCDR/2013/6460.3462] [PMID: 24298468]

[11] Ranjbar R, Sami M. Genetic investigation of beta-lactam associated antibiotic resistance among *Escherichia Coli* strains isolated from water sources. Open Microbiol J 2017; 11: 203-10.
[http://dx.doi.org/10.2174/1874285801711010203] [PMID: 29151997]

[12] Tissera S, Lee SM. Isolation of extended spectrum β-lactamase (ESBL) producing bacteria from urban surface waters in Malaysia. Malays J Med Sci 2013; 20(3): 14-22.
[PMID: 23966820]

[13] Nicolaou KC, Chen JS, Edmonds DJ, Estrada AA. Recent advances in the chemistry and biology of naturally occurring antibiotics. Angew Chem Int Ed Engl 2009; 48(4): 660-719.
[http://dx.doi.org/10.1002/anie.200801695] [PMID: 19130444]

[14] Calderon CB, Sabundayo BP. Antimicrobial classifications: drugs for bugs. In: Schwalbe R, Steele-Moore L, Goodwin AC. Antimicrobial Susceptibility Testing Protocols.Boca Raton, London, New York: CRC Press, Taylor and Francis Group 2007; pp. 7-48.
[http://dx.doi.org/10.1201/9781420014495.ch2]

[15] Etebu E, Arikekpar I. Antibiotics: Classification and mechanisms of action with emphasis on molecular perspectives. Int J Appl Microbiol Biotechnol Res 2016; 4: 90-101.

[16] Dowling A, Dwyer JO, Adley CC. Antibiotics: mode of action and mechanisms of resistance. Antimicrobial Research: Novel bioknowledge and educational programs. Formatex Research Center; Badajoz, Spain: 2017.

[17] Kaur J, Chopra S, Sheevani , Mahajan G. Modified double disc synergy test to detect esbl production in urinary isolates of *Escherichia coli* and *Klebsiella pneumoniae*. J Clin Diagn Res 2013; 7(2): 229-33.
[http://dx.doi.org/10.7860/JCDR/2013/4619.2734] [PMID: 23543257]

[18] Ojdana D, Sacha P, Wieczorek P, *et al.* The occurrence of $bla_{CTX-M}$, $bla_{SHV}$ and $bla_{TEM}$ genes in extended-spectrum β-lactamase-positive strains of *Klebsiella pneumoniae*, *Escherichia coli*, and *Proteus mirabilis* in Poland. Int J Antibiot 2014: 1-7.

[19] Costa PM, Loureiro K, Matos AAJF. Transfer of multidrug-resistant bacteria between itermingled ecological niches: the interface between humans, animals and the environment. Int J Environ Res Public Health 10(1): 276-94.

[20] Magiorakos AP, Srinivasan A, Carey RB, *et al.* Multidrug-resistant, extensively drug-resistant and pandrug-resistant bacteria: an international expert proposal for interim standard definitions for acquired resistance. Clin Microbiol Infect 2012; 18(3): 268-81.
[http://dx.doi.org/10.1111/j.1469-0691.2011.03570.x] [PMID: 21793988]

[21] Lu SY, Zhang YL, Geng SN, *et al.* High diversity of extended-spectrum beta-lactamase-producing bacteria in an urban river sediment habitat. Appl Environ Microbiol 2010; 76(17): 5972-6.
[http://dx.doi.org/10.1128/AEM.00711-10] [PMID: 20639374]

[22]   Tekiner IH, Özpınar H. Occurrence and characteristics of extended spectrum beta-lactamase-
       -producing Enterobacteriaceae from foods of animal origin. Braz J Microbiol 2016; 47(2): 444-51.
       [http://dx.doi.org/10.1016/j.bjm.2015.11.034] [PMID: 26991276]

[23]   Ahmed OI, El-Hady SA, Ahmed TM, Ahmed IZ. Detection of bla SHV and bla CTX-M genes in
       ESBL producing *Klebsiella pneumonia* isolated from Egyptian patients with suspected nosocomial
       infections. Egypt J Med Hum Genet 2013; 14: 277-83.
       [http://dx.doi.org/10.1016/j.ejmhg.2013.05.002]

[24]   Slekovec C, Plantin J, Cholley P, *et al.* Tracking down antibiotic-resistant *Pseudomonas aeruginosa*
       isolates in a wastewater network. PLoS One 2012; 7(12): e49300.
       [http://dx.doi.org/10.1371/journal.pone.0049300] [PMID: 23284623]

[25]   Yu CY, Ang GY, Ngeow YF, Tee KK, Yin WF, Chan KG. Genome sequences of two multidrug-
       resistant *Proteus mirabilis* strains harboring CTX-M-65 isolated from Malaysia. Genome Announc
       2016; 4(6): e01301-16.
       [http://dx.doi.org/10.1128/genomeA.01301-16] [PMID: 27856593]

[26]   Endimiani A, Luzzaro F, Brigante G, *et al. Proteus mirabilis* bloodstream infections: risk factors and
       treatment outcome related to the expression of extended-spectrum β-lactamases. Antimicrob Agents
       Chemother 2005; 49(7): 2598-605.
       [http://dx.doi.org/10.1128/AAC.49.7.2598-2605.2005] [PMID: 15980325]

[27]   Barbosa EGV, Aburjaile FF, Ramos RTJ, *et al.* Value of a newly sequenced bacterial genome. World J
       Biol Chem 2014; 5(2): 161-8.
       [PMID: 24921006]

[28]   Dark MJ. Whole-genome sequencing in bacteriology: state of the art. Infect Drug Resist 2013; 6: 115-
       23.
       [http://dx.doi.org/10.2147/IDR.S35710] [PMID: 24143115]

[29]   Dastmalchi S, Hamzeh-Mivehroud M, Sokouti B. Applied case studies and solutions in molecular
       docking-based drug design. Hershey, PA: Medial Information Science Reference 2016.
       https://books.google.com.my/
       [http://dx.doi.org/10.4018/978-1-5225-0362-0]

[30]   Ferreira LG, Dos Santos RN, Oliva G, Andricopulo AD. Molecular docking and structure-based drug
       design strategies. Molecules 2015; 20(7): 13384-421.
       [http://dx.doi.org/10.3390/molecules200713384] [PMID: 26205061]

# CHAPTER 11

# Exploitation of *Cryptococcus neoformans* Isolated from Cow Dung in Bioremediation of Medical Waste

**S. Sumathi**[*], **R. Akshaya** and **R. Padma**

*Department of Biochemistry, Biotechnology and Bioinformatics, Avinashilingam Institute for Home Science and Higher Education for Women, Coimbatore – 641 043, Tamil Nadu, India*

**Abstract:** Hospital plays a critical role in prevention and treatment of diseases, as well as rehabilitation and improvement of public health. The biomedical waste is the waste that is generated during the diagnosis, treatment or immunization of human beings or animals or in the production or testing of biological components. The application of proper waste management techniques and their identification help improve the understanding of a good waste management practice. A cost-effective environment-friendly technology is also required. The present study was carried out to find the significance and to recommend the development method of bioremediation using fungal strains isolated from cow dung. Further, the isolated fungal strain was identified by using the 18S r-RNA sequencing method, which helps to identify the species. 18S r-RNA sequencing is one of the techniques that have been used to identify microorganisms. Thus this strain was identified as *C. neoformans* based on its morphological characteristics and the r-DNA sequencing of its region data, followed by FESEM characterization. The surface texture and morphology of medical wastes like catheter, IV tubes, were analysed. The fungal strains grown in the biomedical waste powder were observed for their ability to grow and absorb the waste. The results showed that *Cryptococcus neoformans* strongly adheres to the surface of polyethylene powder, indicating the utilization of medical waste. From the results obtained, it was clear that *C. neoformans* has the ability to degrade medical wastes like catheter and IV tubes. Bioremediation of the medical wastes can reduce pollutants in the soil and atmosphere and cause less harmful effects to the plants and animals.

**Keywords:** Biodegradtion of Plactics, Biomedical Waste, Bioremediation, Bollution.

## INTRODUCTION

The waste produced during the diagnosis, treatment, immunization of human and

[*] **Corresponding author S. Sumathi:** Department of Biochemistry, Biotechnology and Bioinformatics, Avinashilingam Institute for Home Science and Higher Education for Women, Coimbatore – 641 043, Tamil Nadu, India; Tel: 9843021733; E-mail: sumii.venkat@gmail.com

**J. Senthil Kumar, P. Ponmurugan & A. Vinothkanna (Eds.)**
All rights reserved-© 2020 Bentham Science Publishers

animal research activities in the production or testing of biological samples or in health camps, is referred to as biomedical waste. Biomedical waste is an extremely ultra-hazardous type of waste; if it is not managed properly, it can lead to chronic health and environmental problems [1]. Waste refers to any useless, unwanted, rejected substance or material which cannot be used further. This includes any substance that is spilled, leaked, pumped, poured, emitted or dumped onto the land, water and air. Biomedical waste generated in the hospitals falls under two major categories, non-hazardous waste and bio-hazardous waste [2]. Non- hazardous waste includes non-infected plastic, cardboard, packaging material, paper, whereas biohazardous waste may be infectious wastes such as sharps, non-sharps, plastics, liquid wastes, *etc.*, or non-infectious wastes that include radioactive waste, discarded glass, chemical waste, cytotoxic waste, incinerated waste, *etc* [3]. Waste generated by health care activities also includes cultures, infectious agents, human tissues, organs, body parts or blood, used and unused sharp objects such as broken glassware and lancets that have been used to puncture or cut the body, as well as body fluids or wastes [4]. The waste generated by the hospitals is a major threat to the person handling those wastes. It can be injurious to humans or animals and deleterious to the environment. When the hazardous and non-hazardous wastes are mixed together, then the whole waste becomes harmful. The microbes present in the waste can leach out and contaminate the environment [5]. Different sources of liquid waste in the hospital include waste disposed from operation theatres, laboratories of microbiology, biochemistry, histopathology, radiology, blood bank, *etc.* Health care waste involves both organic and inorganic substances that enhance the growth of pathogenic microorganisms [6]. If hospital waste is dumped into the municipal sewage, it may create health issues and imbalance in the microbial community present in the sewage systems, which in turn affects the biological treatment process. It is important to be aware of the sources of waste and treat the waste before it is dumped into the environment [7]. The indiscriminate dumping of medical wastes by hospitals and nursing homes is a source of pollution that poses threat to the health and surroundings. In order to overcome this crisis, the BioMedical Waste (Handling and Management) Rules were proposed in July 1998. The rules highlighted medical waste disposal practices in India. The emphasis is on ensuring a process change that will enable health care facilities to handle their waste through proper training and capacity building [8]. These rules are applicable to all or any persons, who generate, collect, receive, store, transport, treat, dispose or handle biomedical wastes. Hospitals, nursing homes, clinics, dispensaries, veterinary institutions, animal homes, pathological laboratories and blood banks come under this category. Although the bio-medical waste (Management and Handling) rules have already been introduced a few years back, not much attention has been paid to manage them [9]. Biomedical waste is hazardous since it has an inherent potential for dissemination of infection, both nosocomial within health care settings as well as to persons working outside health care facilities, like waste handlers, scavenging workers and also to the general public. It is reported that 60% of all hospital staff sustain injuries from sharps during various procedures undertaken in health care facilities. Cytotoxic and chemical waste is mutagenic and teratogenic. Additional hazard includes recycling of disposables without being even washed. Safe disposal of biomedical waste is also a legal requirement in India. The objectives of BMW management are:

- To stop transmission and spreading of pathogens and diseases.
- To prevent injury to individuals in health care services and workers who handle BMW.
- To stop general exposure to the cytotoxic, genotoxic and chemical biomedical wastes due to their harmful effects.
- To prevent environmental degradation.

The currently adopted procedures to treat biomedical waste include chemical process, thermal process, mechanical process, irradiation process and biological process. Chemical disinfectants are usually used for killing microorganisms and inactivating hazardous pathogens [2]. Chemical processes use chemicals such as sodium hypochlorite, hydrogen peroxide, chlorine, *etc.*, that act as disinfectants for biomedical waste. The thermal process utilizes heat to disinfect [9]. Autoclaving is a low heat thermal method that uses steam for treating medical waste. It is commonly used for the human body fluid waste, sharps, and microbiology laboratory waste. This system requires extreme temperature (thermal) that produces steam to decontaminate medical waste [10]. Microwave is an emerging technology to treat biohazardous waste, including material from healthcare facilities [11]. It potentially kills the pathogens present in the biomedical waste [12]. Incineration is a common method used to treat health care waste that uses thermal decomposition at high temperatures between 900 and 1200°C to destroy the organic fraction of the waste. Compaction and shredding constitute the mechanical process that is carried out to reduce the quantity of waste and also to destroy plastic and paper waste [13]. Reverse polymerization uses microwave energy to break down complex molecules, and in this way, to treat medical waste. The irradiation process exposes wastes to ultraviolet radiation in an enclosed chamber. The wave can generate heat to treat waste materials and kill all the microorganisms [10]. The waste materials completely lose their physical form and reduce in volume [11]. Hot air ovens have been used to sterilise glassware and other reusable instruments and infectious health waste [14]. Gas and vapor sterilization employs gaseous or vaporized chemicals as the sterilizing agents. Ethylene oxide is the most commonly used agent [15]. Biological process uses biological enzymes for treating medical waste. It is very clear that biological reactions will not only clean the waste but also cause the destruction of all the organic constituents so that only plastics, glass, and other inert remain within the residues [16]. Proper medical waste management is the mainstay of hospital cleanliness, hospital hygiene and maintenance activities. Appropriate hospital waste management system is an important element of quality assurance in hospitals. The ultimate aim of waste management is the prevention of disease and protection of the environment [17]. It is the moral duty of health care workers to prevent hospitals from becoming centers of disease rather than the center of cure. Awareness should be created among the public regarding the health and environment hazards associated with inappropriate segregation, collection, storage, transport, handling, treatment and disposal of health care waste. Regular training program for all the sections of health care workers with special emphasis on waste handlers is essential [18]. Identifying safe, efficient, sustainable economic and culturally acceptable waste management practices and technologies and enabling the participants to spot the systems according to their specific needs is an important area to be focused [5].

Cow dung, excreta of a bovine animal, is a cheap and easily available bioresource on our planet. Many traditional uses of cow dung, such as for burning of fuel, as mosquito repellent and cleansing agent, are already known in India [19]. Synthetic and semi-synthetic pharmaceuticals are known to pollute the aquatic, terrestrial, and atmospheric environment. Cow dung is the most important source of bio-fertilizer, but at the same time, cow's urine, cow's horn and the dead body of a cow can be used for preparing an effective bio-fertilizer. Cow dung and cow's excreta product are being used as fertilizers and pest repellent, respectively, in agricultural practice [20]. Cow dung is a very effective manure for reducing the bacterial and fungal pathogenic disease. Cow dung slurry consists of bacteria, fungi and actinomycetes [21].

It showed a positive response in the suppression of mycelia growth of plant pathogenic fungi like *Fusarium solani, Fusarium oxysporum* and *Sclerotinia sclerotiorum*. Therefore, the application of cow dung is a proper and sustainable way to enhance not only the productivity of yield but also minimize the possibilities of disease [20]. Cow dung is not a waste material, but it is a purifier of all wastes present in the nature. *Periconiella* species of fungus isolated from cow dung was found to be an excellent degrader of plastics. It was found to be a cheap, safe, and environment-friendly method of medical waste disposal [22]. Biomedical wastes, especially catheter, IV tubes, are made up of polyethylene, Teflon, latex, polyurethane and thermoplastic elastomers. Biomedical wastes like IV tubes, catheters are disposed of without prior treatment, which may lead to the spread of infections [23]. Benzene is one of the compounds released from municipal sludge, which is carcinogenic and is not bactericidal in nature. Bioremediation of benzene can be done using cow dung microflora.

*Pseudomonas putida* is a potential benzene degrader isolated from the cow dung microflora and it has the ability to degrade benzene at various time intervals. *Pseudomonas plecoglossicida* is a novel organism for bioremediation of hazardous compounds like cypermethrin and chlorpyrifos by *Pseudomonas aeruginosa*. These microorganisms obtained from cow dung have the ability to perform bioremediation in laboratory setups, and these can also be applied to pesticide-contaminated soil and water. Cow dung slurry can also be effectively used for degrading phenol and is also used as a source of microbial consortium for bioremediation of fenvalerate soil. It converts the toxic materials into nutrient, biomass and carbon dioxide through biodegradation [24].

With this background, the present study was formulated with the following objectives:

a. To isolate the microbial species from cow dung.
b. To characterize the microbial species and identify the species.
c. To evaluate the effect of the identified species in treating the biomedical waste.
d. To check the efficiency of biodegradation of biomedical waste using microbes.

## MATERIALS AND METHODS

Biomedical waste (BMW) generated in our country on a day-to-day basis is massive and contains infectious and hazardous materials [25]. Bio-medical waste means any solid or liquid waste, which is generated during the diagnosis, treatment or immunization of human beings or animals or in research [26]. A major problem with current biomedical waste management in several hospitals is that the application of biowaste regulation is inadequate as some hospitals dispose of waste in an improper and indiscriminate manner [27]. Various communicable diseases, which spread through water, sweat, blood, body fluids and contaminated organs, need to be prevented. The recycling of disposable syringes, needles, IV sets and glass bottles without proper sterilization is responsible for the spread of Hepatitis, HIV, and other viral diseases. It is the primary responsibility of health administrators to control hospital waste in the most safe and eco-friendly manner [28]. The hospital waste, including body elements, organs, tissues, blood and body fluids with soiled linen, cotton, bandage and plaster casts from infected and contaminated areas, should be properly collected, segregated, stored, transported, treated and disposed of in a safe manner to prevent healthcare facility or hospital from spreading non-heritable infections. Cow dung has been traditionally used as organic fertilizer in India for centuries. The addition of cow dung increases the mineral status of soil, enhancing the resistance of plant against pests and diseases. As per Ayurveda, it may also act as a purifier for all the wastes that exist within the nature. *Periconiella* species of fungus isolated from cow dung was found to be good degrader of biomedical waste. Nowadays, there is an increasing research interest in developing the applications of cow dung microorganisms for the management of environmental pollutants [29].

## Sample Collection

The fresh cow dung samples were collected from the farm and preserved in sterile polythene bags and stored in a refrigerator till use. The isolation of microbes was then carried out. Microbial analysis of cow dung sample involved a series of sequential dilutions used to reduce the dense culture of cells to a more usable concentration. Each dilution will reduce the concentration of microbial species by a specific amount. Isolation and enumeration of fungi from the cow dung sample were done by the agar plate method. The medium used for the isolation of fungi was potato dextrose agar [30]. By using the serial dilution method, cow dung samples were prepared. Following the serial dilution technique, the samples were preserved and labeled. Out of 10 dilutions, 0.1ml of the last three dilutions were taken for spreading.

## Isolation and Screening of Fungal Strains from Cow Dung Sample

Fungus was isolated from cow dung sample using Potato dextrose agar. The fungal colonies were isolated, purified and stained using lactophenol staining.

## Primary Screening of Fungal Strains using Morphology Characterization

Fungal species are usually identified on the basis of their morphological characteristics. Colony morphology is a method used to describe the characteristics of an individual colony of fungi growing on agar in a petri dish. The morphologies are observable through an electron microscope. The morphologies are most important for the identification of organisms. Colony morphology was observed after 5 days of incubation. Fungi produce different looking colonies. Few colonies may be coloured, some circular in shape and others irregular. Colony morphology includes form, size, elevation, margin or border, surface, opacity, pigmentation, mycelium and spore. The shape of the colony may vary from circular, irregular, filamentous or rhizoid. The size of the colony varies from large to small and it is measured using a ruler. Tiny colonies are referred to as punctiform. The magnified edge shape of the colony is observed using a microscope. The margin of the colony varies from entire, undulate, filiform, curled or lobate. The surface of the colony varies from smooth, rough, glistening, wrinkled or dull. The elevation of the colony describes the side view of the colony. Elevation of the colony varies from being raised, corvex, flat and crateriform. That is, the colony rising above the agar can be determined by its height. Opacity of the colony determines whether the colony is transparent, opaque or translucent. Pigmentation of the colonies involves observing the color of the colonies such as white, buff, red, purple, *etc*. Mycelium is the thread-like structure found in fungi, which is more visible to the naked eye. The spores may be microscopic, tough and resistant bodies that are round in shape.

## Identification of Fungal Strains using Lacto Phenol Staining

Lacto phenol cotton blue staining is the most widely used method for the characterization of fungi. Lacto phenol cotton blue stain is formulated with lactophenol and cotton blue. The slides were observed under the light microscope for the identification of fungi [30].

## Fungal Identification using 18S r-RNA Sequencing

18SrRNA sequencing is a common sequencing method used to identify and compare the microorganisms present within the given sample. It is one of the techniques that have been used recently to identify microorganisms, including those that are capable of degrading plastics. This technique is used to identify the genus and species name of an organism. It is a detailed procedure that is carried out to identify the microorganisms.

## Screening of Medical Waste Degrading Fungi

Once the fungal strain was identified, we planned to test the ability of the fungus to degrade biomedical waste. We chose catheters and IV tubes that are usually disposed of without prior treatment. This may, in turn, pollute the soil as well as cause the spread of pathogens. So the fungus was used to treat the biomedical waste. The fungal species which showed the ability to degrade medical waste were segregated from other fungal isolates, which did not exhibit the ability to deteriorate medical waste. The assay was carried out to screen the biomedical waste degrading fungi. The surface morphology was analyzed through Field Emission Scanning Electron Microscopy (FESEM) to evaluate the structural changes in isolated fungal species after incubation with biomedical waste. The biomedical waste, namely catheter tubes, IV tubes, were collected from hospitals and powdered [31]. The fungal species were grown in different concentrations of medical waste powder. The changes in the morphology were observed using FESEM.

## RESULTS

Microorganisms such as bacteria, algae, yeast and fungi have the potential to degrade plastics. Indigenous microorganisms with specific metabolic capabilities play a significant role in the biodegradation of wastes. Biological treatment is preferred over physicochemical processes due to its feasibility and efficacy. Cow dung is an eco-friendly source of organic matter. The fungi that germinate, grow and sporulate on dung are termed as 'Coprophilous'. The word Coprophilous literally means "dung loving". The nitrogenous compounds present in the cow dung influence the growth of fungi [4]. Biomedical waste is dangerous since it contains pathogenic bacteria, viruses, toxic chemicals, mould and radioactive materials. This kind of waste could contaminate other waste and is also infectious. The studies on the degradation of biomedical waste are much rarer than those on other types of waste, most likely due to a high risk of danger in handling samples of biomedical waste. However, research on biomedical waste bioremediation should be continuously carried out because the increasing number of hospitals require more effective and efficient biomedical waste treatments for the safe disposal of waste material [32]. In the present study, the fungus was isolated from cow dung to treat the biomedical waste and identified by the sequencing technique as *C. neoformans*. Fungal identification was done using 18S r-RNA sequencing to determine the genus and species name of the organism. The structural changes in isolated fungal species were analysed through FESEM, followed by screening of the medical waste degrading fungi by Plate assay method. Serial dilution was done to estimate the concentration of unknown sample by counting the number of colonies cultured in the petri dishes. By knowing the aliquot volume, colony counts were converted into concentrations (CFU/ml). Out of 10 dilutions, 0.1ml of the last three dilutions ($10^{-6}$, $10^{-8}$ and $10^{-9}$) were taken for spreading. The dilution factor $10^{-6}$ contained three colonies, dilution factor $10^{-8}$ contained four colonies and the dilution factor $10^{-9}$ contained three colonies cultured in the petri dishes. The viable cell counts of the Colony forming units (CFU) averaged $3 \times 10^{-5}$ CFU/ml, $4 \times 10^{-10}$ CFU/ml and $3 \times 10^{-11}$ .Olawepo *et al.* (2018) reported that fungal species identified in the cow dung were *Aspergillus niger, Aspergillus flavus, Penicillium chrysogenum and Neurospora crazza*. Among the fungi, *Penicillium* and *Aspergillus* species are capable of performing bioremediation of hydrocarbon. Total fungi counts were determined for the 95 cocoa bean samples by the pour plating technique. The growth of the fungal colonies was consistent on PDA media inoculated with the dilution factors $10^{1}$ and $10^{3}$, as reported by the study [33]. Isolation of fungal strains by serial dilution is a common method. Hence we too employed this method to isolate the fungus.

Fungal species are usually identified on the basis of their morphology characteristics. The colony appearance of each fungal isolate was characterized on Potato dextrose agar media. Microscopic characteristics of the fungal isolates were examined using an electron microscope. The morphologies are most important for the identification of organisms. Lacto phenol cotton blue stain was used to identify the isolated microorganism. Slides mounted with cotton blue dye stained the spores of fungal species in blue. The results showed that LPCB stained the cell wall intensely blue. Staining results were confirmed by 18S r-RNA sequencing. The structural changes in isolated fungal species were analysed through FESEM, followed by the screening of the medical waste degrading fungi by the Plate assay method.

## PRIMARY SCREENING OF FUNGAL STRAINS USING MORPHOLOGY CHARACTERIZATION

Fungal species are usually identified on the basis of their morphology characteristics. The colony appearance of each fungal isolate was characterized on Potato dextrose agar media. Microscopic characteristics of the fungal isolates were examined using an electron microscope. The morphologies are most important for the identification of organisms. Lacto phenol cotton blue stain was used to identify the isolated microorganism. Slides mounted with cotton blue dye stained the spores of fungal species in blue. The results showed that LPCB stained the cell wall intensely blue. Staining results were confirmed by 18S r-RNA sequencing (Table **1**).

**Table 1. Colony Characterization.**

| Dilution | Size | Shape | Margin | Diffusion in Media | Pigmentation | Mycelium | Spore |
|---|---|---|---|---|---|---|---|
| $10^{-6}$ | Small | Circular | Undulate | Raised | Black | - | - |
| $10^{-6}$ | Moderate | Circular | Entire | Flat | Buff | - | - |
| $10^{-6}$ | Large | Irregular | Entire | Raised | Buff | - | Yes |
| $10^{-9}$ | Moderate | Circular | Undulate | Raised | Buff | - | - |
| $10^{-9}$ | Small | Irregular | Undulate | Flat | Black | - | - |
| $10^{-9}$ | Moderate | Circular | Entire | Flat | Buff | - | - |
| $10^{-9}$ | Small | Circular | Entire | Flat | White | - | - |
| $10^{-8}$ | Large | Circular | Undulate | Raised | Black | - | Yes |
| $10^{-8}$ | Small | Irregular | Undulate | Flat | Buff | - | - |
| $10^{-8}$ | Small | Irregular | Undulate | Flat | White | - | - |

## Molecular Identification of Fungal Strains

The isolated fungal species were subjected to 18S r-RNA sequencing. 18S r-RNA sequencing is one of the techniques that have been used to identify microorganisms. It is used to identify genus and species name of an organism. It helps in the classification and quantitation of microbes within a complex biological mixture. Positive samples of 18S r-RNA PCR were selected for further species confirmation by DNA sequencing. Primers ITS1 and ITS4 were used to amplify the region by PCR. The PCR products were sequenced using a DNA sequencer and analyzed with the BLAST program provided by the National Centre for Biotechnology Information (NCBI) to confirm the fungal species.

### Phylogeny Analysis

Fig. (**1**) shows the phylogenic analysis. The purified product was sequenced using ITS1 (5' TCCGTAGGTGAACCTGCGG 3') and ITS4 (5' TCCTCCGCTTATTGATATGC 3'). The gel image of the PCR product is shown in Fig. (**2**). Sequencing reactions were performed using ABI PRISM® BigDyeTM Terminator Cycle Sequencing Kits with AmpliTaq® DNA polymerase (FS enzyme) (Applied Biosystems). Single-pass sequencing was performed on each template using below 16s rRNA universal primers. The fluorescent-labeled fragments were purified from the unincorporated terminators. The samples were resuspended in distilled water and subjected to electrophoresis in an ABI 3730xl sequencer (Applied Biosystems).

**Fig. (1).** Phylogenic analysis.

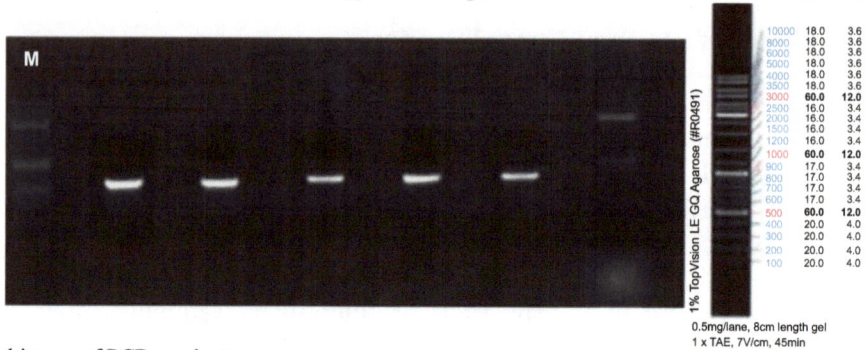

**Fig. (2).** Gel image of PCR product.

SEQUENCE

DATA IN FASTA FORMAT >Contig

GAGAATATTGGACTTTGGACCATTTATCTACCCATCTACACCTGTGAA
CTGTTTATGTGCTTCGGCACGT
TTTAGACAAACTTCTAAATGTAGTGAATGTAATCATATTATAACAATA
ATAAAACTTTCAACAACGGATC
TCTTGGCTTCCACATCGATGAAGAACGCAGCGAAATGCGATAAGTAAT
GTGAATTGCAGAATTCAGTGAA
TCATCGAGTCTTTGAACGCAACTTGCGCCCTTTGGTATTCCGAAGGGC
ATGCCTGTTTGAGAGTCATGAA
AATCTCAATCCCTCGGGTTTTATTACCTGTTGGACTTGGATTTGGGTGT
TTGCCGCGACCTGCAAAGGAC
GTCGGCTCGCCTTAAATGTGTAAGTGGGAAGGTGATTACCTGTCAGCC
CGGCGTAATAAGTTTCGCTGGG
CCAATGGGGTAGTCTTCGGCTTGCTGATAACAACCATCTCTTTTTGT

**Organism**

***Cryptococcus Neoformans***

The 18s rRNA sequence was identified using the NCBI blast similarity search tool. The phylogeny analysis of query sequence with the closely related sequence of blast results was performed, followed by multiple sequence alignment. MUSCLE 3.7 program was used for multiple alignments of sequences. Finally, the program PhyML 3.0 aLRT was used for phylogeny analysis and HKY85 as a substitution model.

**FESEM CHARACTERIZATION OF FUNGAL STRAIN**

Field Emission Scanning Electron Microscopy was used to examine the topography of fungal species (*Cryptococcus neoformans*) treated with medical waste. FESEM provides the images at 100 and 10nm scales. The images were acquired using FESEM and the surface texture and morphology of medical wastes like catheter, IV tubes, were analysed to check for any structural changes after incubating for 5 days, as shown in Fig. (**3**).

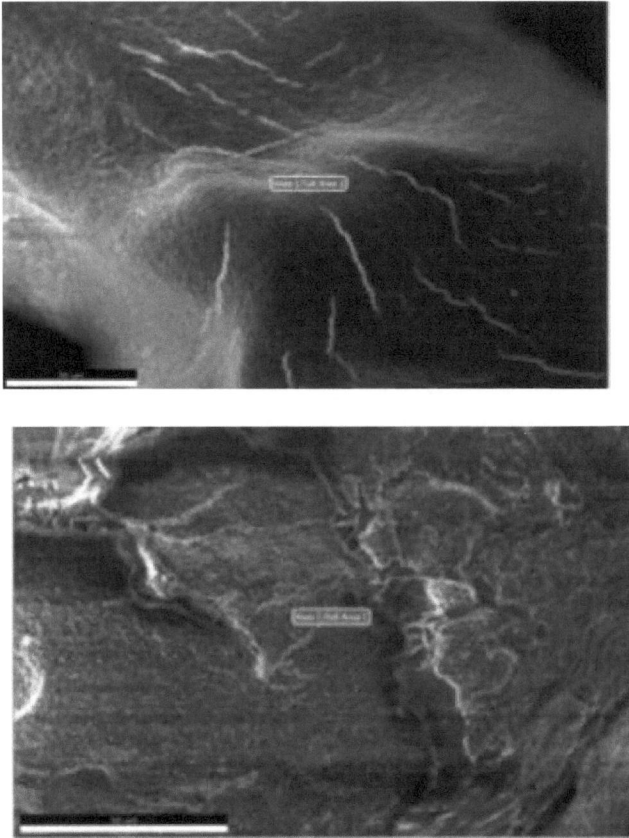

**Fig.(3).** FESEM analysis of *C. neoformans.*

## FESEM ANALYSIS OF *C. NEOFORMANS* TREATED WITH MEDICAL WASTE

A typical topographic FESEM image represented the whole surface morphology of *C. neoformans* before and after treatment with medical waste.

Different quantities of biomedical waste were grown along with the inoculated fungi. Among these, 0.02g of medical waste treated with fungal species showed a better result. So it was chosen for FESEM analysis to examine the morphological changes before and after treatment with medical waste.

FESEM images were captured at different magnifications of 5μm, 10μm, 15μm, 25μm and 50μm. The clear image was observed at 50μm magnification showing structural changes in the fungal strain with treated and untreated medical waste. The image showed that *C. neoformans* strongly adhered to the surface of polyethylene powder, indicating the utilization of medical waste. The results of FESEM analysis showed that *C. neoformans* possessed the ability to deteriorate medical wastes like catheter, IV tubes, *etc.*

## CONCLUSION

Hospitals play a critical role in prevention and treatment of diseases, as well as rehabilitation and improvement of public health. The biomedical waste gets accumulated as a result of procedures related to diagnosis, treatment or immunization of human beings or animals or in production or testing of biological components. The application of proper waste management techniques and their identification help to improve the understanding of a good biomedical waste management practice. We also are in the dire need of a cost-effective environment-friendly technology. The present study was carried out to find the significance of and to recommend the development method of bioremediation using fungal strains isolated from cow dung. In the present study, cow dung was found to be an excellent degrader of biomedical waste. Cow dung can be employed to purify all wastes found in the nature. The fungal strain *Cryptococcus neoformans* isolated from cow dung has the ability to degrade biomedical waste. *C. neoformans* causes the majority of infections found in immuno-compromised individuals. Cow dung samples were prepared by using a serial dilution method. The different fungal cultures were purified by using the spread plate method on potato dextrose agar medium. The plates were preserved and labeled for the growth of microorganisms at 37 °C for 5 days. The unique colonies were further sub-cultured for screening and identification of fungal strains. By knowing the aliquot volume, colony counts were converted into concentrations (CFU/ml). Out of 10 dilutions, 0.1ml of the last three dilutions ($10^{-9}$, $10^{-8}$ and $10^{-6}$) represented the viable cell counts averaged $4 \times 10^{-11}$ CFU/ml, $3 \times 10^{-10}$ CFU/ml and $3 \times 10^{-5}$. Based on their morphological characteristics, fungal species have been identified and examined using an electron microscope. From the observations, it was clear that the size of the fungal colony observed was small, moderate and large, and it was measured using a ruler. The shape of the colony observed was circular and irregular. The magnified edges of the colony were observed as entire and undulate. The fungal colony was raised above the agar as well as diffused in media. The pigmentation of the colonies was observed as black and buff. The spores were observed on the colony, but no mycelium was observed. Among the colonies from different dilutions, these results were noticed in the dilution factor $10^{-6}$. The size of the colony identified was moderate and small. The shape of the fungal colony observed was circular and irregular. The margin of the colony observed was entire and undulate. The fungal colony observed was diffused in the agar and few colonies were raised above the media. The colour of the colonies was observed as buff, black and white. Mycelium was not observed; spores appeared on the single colony. No mycelium or spore formation was observed. These results were noticed in the dilution factor $10^{-9}$. The size of the fungal colony observed was large and small. The shape of the colony identified was circular and irregular. The magnified edge of the colony was observed as undulate in all the three colonies. The fungal colony observed was raised and flat in the agar. The colour of the colonies was observed as buff, black and white. Mycelium was not observed; spores appeared on the single colony. Among the colonies from different dilutions, these results were noticed in the dilution factor $10^{-8}$. Microscopic observation of fungal strains using lacto phenol cotton blue staining was done and the results showed that the LPCB stained the cell wall intensely blue. The observations showed a clear photographic image of the fungal strains confirming that the conidiophores were heavy-walled, uncoloured and coarsely rough. Further, the isolated fungal strain was identified by using the 18S r-RNA sequencing method, which helped to identify the species. 18S r-RNA sequencing is one of the techniques that have been used to identify microorganisms. It was used to identify the genus and species name of an organism. 18S r-RNA PCR positive samples were selected for further species confirmation by DNA sequencing. The fungal DNA was isolated and amplified using universal primers ITS1 (5' TCCGTAGGTGAACCTGCGG 3') and ITS4 (5' TCCTCCGCTTATTGATATGC 3'). Amplification was carried out in PCR programmed with an initial denaturation at 94 °C for 3 minutes, followed by 30 cycles at 94°C for 30 seconds (Denaturation), 60°C for 30 seconds (Annealing), 72°C for 1 minute (Extension), and a final extension at 72°C for 10 minutes. PCR product was electrophoresed in 1% agarose gel. The 18s rRNA sequence was identified using the NCBI blast similarity search tool. The phylogeny analysis of query sequence with the closely related sequence of blast results was performed, followed by multiple sequence alignment. MUSCLE 3.7 program was used for multiple alignments of sequences. Finally, the program PhyML 3.0 aLRT was used for phylogeny analysis and HKY85 as a substitution model. 18S r-RNA sequencing results confirmed that the organism belonged to genus *Cryptococcus* and species *C. neoformans*. Thus this strain was identified as *C. neoformans* based on its morphological characteristics and the r-DNA sequencing of its region data. This tree was constructed on the basis of the r-DNA sequence (ITS1 and ITS4) by neighbor-joining method. In our study, fungal strains were isolated and identified using 18S r-RNA sequencing, followed by FESEM characterization. The surface texture and morphology of medical wastes like catheter, IV tubes, were analysed. The fungal strains grown in the biomedical waste powder were noted for their ability to grow and absorb the waste. The results showed that *Cryptococcus neoformans* strongly adhered to the surface of polyethylene powder, indicating the utilization of medical waste. From the results obtained, it was clear that *C. neoformans* has the ability to degrade medical wastes like catheter and IV tubes. The results of the present study revealed that the isolated fungal organism from the cow dung sample possess a good potential to deteriorate the medical wastes like catheter, IV tubes, because of the adherence of this powder to fungal strains. Bioremediation of the medical wastes can reduce pollutants in the soil and atmosphere, causing less harmful effects to the plants and animals.

## CONSENT FOR PUBLICATION

Not applicable.

## CONFLICT OF INTEREST

The authors confirm that this chapter contents have no conflict of interest.

## ACKNOWLEDGEMENTS

Declared none.

## REFERENCES

[1]     Yawson P. Assessment of solid waste management in healthcare facilities in the offinso municipality 2015.

[2]     Datta P, Mohi GK, Chander J. Biomedical waste management in India: Critical appraisal. J Lab Physicians 2018; 10(1): 6-14.
        [http://dx.doi.org/10.4103/JLP.JLP_89_17] [PMID: 29403196]

[3]     Pullishery F, Panchmal GS, Siddique S, Abraham A. Awareness, Knowledge and Practices on Bio-medical waste management among health care professionals in Mangalore - A cross sectional study. IAIM 2016; 3: 29-35.

[4]     Zeeshan MF, Al Ibad A, Aziz A, *et al.* Practice and enforcement of national hospital waste management 2005 rules in Pakistan. East Mediterr Health J 2018; 24(5): 443-450.
        [http://dx.doi.org/10.26719/2018.24.5.443]

[5]     Rajan R, Robin DT, Vandanarani M. Biomedical waste management in Ayurveda hospitals–current practices and future prospectives. J Ayurveda Integr Med 2019; 10(3): 214-21.
        [PMID: 29555257]

[6]     Alam I, Alam G, Ayub S, Siddiqui AA. Assessment of Bio-medical Waste Management in Different Hospitals in Aligarh City. In: Kalamdhad A, Singh J, Dhamodharan K (Eds) Advances in Waste Management. Singapore: Springer 2019: pp/ 501-10.

[7]     Radhakrishna L, Nagarajan P. Isolation and preliminary characterization of bacterial from liquid hospital wastes. Int J Pharm Tech Res 2015; 8: 308-14.

[8]     Berihun D, Solomon Y. Preliminary assessment of the status of hospital incineration facilities as a health care waste management practice in Addis Ababa city, Ethiopia. Adv Recycling Waste Manag 2017; 2: 3.

[9]     Hirani DP, Villaitramani KR, Kumbhar SJ. Biomedical waste: An introduction to its management. Int J Innov Res Adv Eng 2014; 1: 82-7.

[10]    Airlina I. Medical Waste Disposal–The Definitive Guide. Biomedical waste solutions 2015.

[11]    Zimmermann K. Microwave as an emerging technology for the treatment of biohazardous waste: A mini-review. Waste Manag Res 2017; 35(5): 471-9.
[http://dx.doi.org/10.1177/0734242X16684385] [PMID: 28148206]

[12]    Voudrias EA. Technology selection for infectious medical waste treatment using the analytic hierarchy process. J Air Waste Manag Assoc 2016; 66(7): 663-72.
[http://dx.doi.org/10.1080/10962247.2016.1162226] [PMID: 26962884]

[13]    Prasad L L, Reddy P V K. A study on waste management practices in private hospitals in khammam district. CLEARIJRCM, 2017; 8(11): 53-7.

[14]    Capoor MR, Bhowmik KT. Current perspectives on biomedical waste management: Rules, conventions and treatment technologies. Indian J Med Microbiol 2017; 35(2): 157-64.
[PMID: 28681801]

[15]    Patan S, Mathur P. Assessment of biomedical waste management in government hospital of Ajmer city-a study. Int J Res Pharm Sci 2015; 5(1) : 6–11.

[16]    Ghasemi MK, Yusuff RB. Advantages and disadvantages of healthcare waste treatment and disposal alternatives: Malaysian Scenario. Pol J Environ Stud 2016; 25(1): 17-25.

[17]    Joshi S C, Diwan V, Tamhankar A J, *et al.*  Staff perception on biomedical or health care waste management: A qualitative study in a rural tertiary care hospital in India. PLoS One. 2015; 10(5): e0128383.

[18]    Kumar R, Somrongthong R, Shaikh BT. Effectiveness of intensive healthcare waste management training model among health professionals at teaching hospitals of Pakistan: a quasi-experimental study. BMC Health Serv Res 2015; 15: 81.
[http://dx.doi.org/10.1186/s12913-015-0758-7] [PMID: 25889451]

[19]    Gupta KK, Aneja KR, Rana D. Current status of cow dung as a bioresource for sustainable development. Bioresour Bioprocess 2016; 3: 28.
[http://dx.doi.org/10.1186/s40643-016-0105-9]

[20]    Raj A, Jhariya MK, Toppo P. Cow dung for eco-friendly and sustainable productive farming. Environ Sci (Ruse) 2014; 3: 201-2.

[21]    Idham I, Sudiarso S, Aini N, Nuraini Y. Isolation and identification on microorganism decomposers of Palu local cow manure of Central Sulawesi, Indonesia. J Degraded Min Lands Manag 2016; 3: 625.
[http://dx.doi.org/10.15243/jdmlm.2016.034.625]

[22]    Prashanthi M, Sundaram R, Jeyaseelan A, Kaliannan T, Eds. Bioremediation and sustainable technologies for cleaner environment.
[http://dx.doi.org/10.1007/978-3-319-48439-6]

[23]    Deb P K, Kokaz S F, Abed S N, Paradkar A, Tekade R K. Pharmaceutical and biomedical applications of polymersIn basic fundamentals of drug delivery 2019; 203-67.

[24]    Godambe T, Fulekar M H. Cow dung bacteria offer an effective bioremediation for hydrocarbon-benzene. IJBTT 2016; 6(3): 13-22.

[25]    Manasi S. Challenges in biomedical waste management in cities: a ward level study of Bangalore. Adv Recycling Waste Manag 2017; 2: 1-8.

[26]    Karmakar N, Datta S, Datta A, Nag K. A cross-sectional study on knowledge, attitude and practice of biomedical waste management by health care personnel in a tertiary care hospital of Agartala, Tripura. Natl J Res Community Med 2016; 5: 189-95.

[27]    Priya DN, Gupta M, Jyothsna TS, Chakradhar B. Current Scenario of Biomedical Waste Management

in India: A Case Study. In waste valorisation and recycling. Singapore: Springer 2019; pp. 147-58.

[28]  Mishra S, Sahu C, Mahananda M R. Study on Knowledge, Attitude and Practices (KAP) of Bio-medical waste management at Veer Surendra Sai Institute of Medical Science and Research (VIMSAR), Burla. IJERMT 2015; 4(11): 41-3.

[29]  Thilagam L, Nayak BK, Nanda A. Studies on the diversity of coprophilous microfungi from hybrid cow dung samples. Int J Pharm Tech Res 2015; 8: 135-8.

[30]  Basava SPR, Jithendra K, Premanadham N, Reddy PS. Efficacy of iodine-glycerol *versus* lacto phenol cotton blue for identification of fungal elements in the clinical laboratory. Int J Curr Microbiol Appl Sci 2016; 5: 536-41.
[http://dx.doi.org/10.20546/ijcmas.2016.511.063]

[31]  Ahsan A, Ashraf M, Ali S, Aslam R. Isolation and screening of polyethylene degrading fungi from solid waste material. J Agric Basic Sci 2016; 1(1): 1-6.

[32]  Ethica SN, Saptaningtyas R, Muchlissin SI, Sabdono A. The development method of bioremediation of hospital biomedical waste using hydrolytic bacteria. Health Technol (Berl) 2018; 8: 239-54.
[http://dx.doi.org/10.1007/s12553-018-0232-8]

[33]  Kutsanedzie FYH, Chen Q, Hassan MM, Yang M, Sun H, Rahman MH. Near infrared system coupled chemometric algorithms for enumeration of total fungi count in cocoa beans neat solution. Food Chem 2018; 240: 231-8.
[http://dx.doi.org/10.1016/j.foodchem.2017.07.117] [PMID: 28946266]

CHAPTER 12

# Biomedical Waste: Safe Disposal and Recycling

**V. Arun**[1,*] and **B. Sathya Priya**[2]

[1] *Department of Biotechnology, Sri Ramachandra Institute of Higher Education and Research (SRIHER) (DU), Chennai - 600 116, Tamil Nadu, India*

[2] *Department of Environmental Sciences, Bharathiar University, Coimbatore - 641 046, Tamil Nadu, India*

**Abstract:** Biomedical waste is generated due to human and research activities related to medical treatments for human health. The various wastes generated include solid, liquid, chemical forms, etc. and are of varied composition like drugs, solvents, body fluids, anatomical parts, food, *etc*. There is a sharp increase in the production of biomedical waste due to population explosion and other human activities. Hence there is a need to understand biomedical waste in its entirety to provide proper management to handle spillages, accidents, and other issues involving biomedical waste. This chapter explains the definition, types, hazards of biomedical wastes and its current management strategies along with various international treaties and policies regarding biomedical wastes.

**Keywords:** 3 R's Strategy, Biomedical Waste, Biomedical Waste Management, Hazardous Waste, Healthcare Waste Management, WHO Guidelines.

## INTRODUCTION

Biomedical waste management refers to managing waste materials that arise from various areas of human activity, like hospitals, clinics, laboratories, homes, *etc*. Hence a biomedical waste (BMW) is either a solid or liquid waste produced during clinical procedure, diagnosis, or therapy for humans in hospitals/ clinics/ laboratories like blood testing, immunization. BMW also includes waste generated in biomedical/biotechnological/pharmaceutical/clinical research laboratories, where animals and human subjects are involved in medical research. With improved health awareness among people due to the internet and media, there is a marked increase in health check-ups done by people in urban and rural areas. Hence there is a steady, moderate to a heavy output of biomedical waste

* **Corresponding author V. Arun:** Department of Biotechnology, Sri Ramachandra Institute of Higher Education and Research (SRIHER) (DU), Chennai - 600 116, Tamil Nadu, India; Tel: 9444918200; E-mail: arun.v@sriramachandra.edu.in

**J. Senthil Kumar, P. Ponmurugan & A. Vinothkanna (Eds.)**

from these places, and without a proper management system, it could result in outbreaks of infectious diseases among humans and domestic animals.

In order to implement proper and sustainable biomedical waste management systems, we need to understand the source, type, toxicity/ pathological effect, and available disposal methods for biomedical wastes. We also need to understand the rules and regulations pertaining to waste management within the country and increase awareness among the employees and the public for a successful management system [1].

## BIOMEDICAL WASTE AND ITS CLASSIFICATION

Biomedical waste predominantly deals with waste generated by the health care system and its ancillary activities like biomedical, clinical or biotechnological laboratories. The healthcare waste could be classified as either hazardous or non-hazardous based on their impact on the environment and humans. The biomedical waste could be grouped into the following categories based on WHO [1]:

1. Hazardous waste (10-25% of total waste)
2. Solid waste (human or animal anatomic parts, tissues, disposable items, plastics like tubes, catheters, syringes, bins)
3. Liquid waste (body fluids, blood, urine)
4. Sharp waste (needles, scalpels, other blades, broken glass, metal plates)
5. Soiled waste (cotton, swabs, bandages, used containers, plasters)
6. Infectious waste (bacteria, fungi, other pathogens, excreta of human or animals)
7. Chemical waste (radioisotopes, expired or spilled medications, solvents, heavy metals)
8. Non-hazardous waste (75-90%) - administrative, cooking/pantry, housekeeping materials

### Source of Biomedical Waste

According to WHO [1], the different sources of biomedical waste could be broadly classified based on the amount and type of wastes into major and minor sources. The major sources are the places where maximum activities related to biomedical waste generation take place like hospitals, laboratories, clinics, research centers. In these places, a variety of wastes ranging from sharps to radioactive wastes are produced in large quantities owing to testing, treatment and recovery of patients or research subjects. Examples of major sources are hospitals, other health care facilities, laboratories, research centers (medical and veterinary), among others. Minor sources of biomedical waste, according to WHO [1], are places where the quantum of waste and its composition are simple (*i.e.*) with absence of radioactive/cytostatic waste, and which do not mainly consist of human or animal body parts and where sharp waste consists mainly of hypodermic needles. Examples of minor sources are dental clinics, nursing homes, first aid centers, cosmetics centers (ear-piercing, tattoo), home treatments, ambulance and funeral services.

## Hazards Associated with BMW

Biomedical waste is a multi-dimensional waste as it includes infectious waste (pathogens), chemical waste (drugs, solvents), toxic waste (genotoxic, cytotoxic, radioactive, heavy metals), sharps waste (needles, scalpels) and hence could pose a variety of health issues to people involved in the BMW management [1]. Doctors, nurses, maintenance personnel, patients, visitors, workers (BMW management, cleaners, laundry, transport) and scavengers are the people who are exposed mostly to these wastes on a daily basis and hence are at maximum risk [2, 3].

## Risk Related to Infectious and Sharp Waste

The presence of pathogens in biomedical waste that are not treated properly could result in disease outbreaks like cholera and dengue. Contaminated needles, scalpels and concentrated pathogen cultures are the major threats to the spread of disease to susceptible host or environment. These pathogens could enter the human body through skin (cut/ wounds), mucous membrane, inhalation or ingestion. The various infections caused include GI infections (*E. coli, Salmonella, and Vibrio cholera*), respiratory infections (*M. tuberculosis, S. pneumoniae*), skin infections (*Streptococcus sp*), viral infections (HIV, Hepatitis, Influenza), multidrug-resistant organisms (MRSA).

## Risks Related to Chemical, Pharmacological and Genotoxic Waste

The chemicals and drugs used in health care are mostly hazardous in nature and the common injuries observed with chemical wastes are burns (physical injury) and intoxication (acute/chronic) due to inhalation. The various hazardous properties of the chemical wastes are toxic, corrosive (acids, alkali), explosive, flammable, reactive and radioactive. Additionally, mercury and disinfectants (chlorine, quaternary ammonium), cytotoxic or genotoxic drugs (carcinogenic or mutagenic) have also been generated in biomedical facilities. More information can be obtained from the International Agency for Research on Cancer (IARC) on the effects of cytotoxic drugs. Yet, there is a need for research and epidemiological surveys on the impact of these wastes on the environment and public.

## Laws and Regulation of BMW Management

Biomedical waste is a global phenomenon, as every country in the world encounters these wastes on a daily basis. Managing these wastes requires active participation of government regulatory bodies, local authorities, hospital/clinic/ university, personnel involved in health care and related departments. For a successful and sustainable BMW management, a national policy is needed that proposes rules and regulations regarding the various aspects of biomedical waste

like production, safe disposal, the responsibility of polluter, *etc.* A national policy is a guideline in concurrence with international treaties and conventions adopted for the safe management of hazardous waste that identifies the needs and problems of the country, and should take into account sustainable growth, public health and environment. Detailed guidelines are available from WHO, UNEP and other non- governmental organizations (NGOs).

WHO recommends the following five guiding principles for framing of rules and regulations for the management of biomedical waste:

1. The polluter principle states that the producers of waste are responsible legally and financially for the safe disposal of the waste and also liable for any damage caused thereof.
2. The precautionary principle was originally defined and adopted under the Rio Declaration on Environment and Development (UNEP, 1992) [4]. This principle provides regulations on the safety and health of people involved in the management of biomedical waste and the environment.
3. The duty of care principle states that the person involved in the handling of hazardous material or equipment or waste is ethically responsible for ensuring the utmost care in that task. In order to achieve this, all the concerned people (production, storage, transport, treatment and disposal) should be trained and authorized to handle hazardous material or its waste.
4. The proximity principle states that the location of the place where the treatment and disposal of hazardous waste take place should be nearest to its source of generation or utilization [1]. This would minimize the risk of transportation of waste and also suggests recycling of wastes wherever possible.
5. The prior informed consent principle pertains to the protection of the environment and public health from hazardous waste [1]. It recommends that the stakeholders should be informed about the hazards and safety precautions regarding the biomedical waste, and consent has to be obtained. In a healthcare facility, personnel involved in the treatment, disposal and transport of biomedical waste have to be informed and sought consent.

## International Treaties and Conventions

Some of the key international treaties and conventions related to healthcare facilities, environmental protection and sustainable development are listed below and should be taken into account for drafting waste management policies.

### The Basel Convention

This is an important treaty that discusses the Control of Trans-Boundary Movements of Hazardous Wastes and their Disposal. The main aim of this treaty is the protection of the global environment and human health against hazardous wastes, and it has 170 countries as members. The movement of hazardous wastes between countries is regulated by the prior informed consent principle and detailed instructions are provided in "Technical guidelines on the environmentally sound management of biomedical and healthcare wastes (Y1; Y3)" [5]. Apart from periodical meetings, the convention is regularly modified based on the decisions made at the Conference of the Parties to the Basel Convention [4 - 8].

### The Stockholm Convention

It is one of the most important treaties that deal with persistent organic pollutants (POP) and their impact on human health and the environment. POPs are highly stable organic compounds and include polychlorinated dibenzo-p-dioxins and dibenzofurans released from medical waste incinerators. The POPs get accumulated in fatty tissues in all living organisms and cause severe damage. The convention in 2006 released guidelines on best available techniques (BAT) and best environmental practices (BEP). According to this convention, source reduction, segregation, resource recovery and recycling, training, proper collection, and transport are some of the best environmental practices. It also states that using BAT, the amount of polychlorinated dibenzo-p-dioxins and dibenzofurans in air emissions and wastewater has to be brought to less than 0.1 ng/L I-TEQ/Nm$^3$ (at 11% $O_2$) and 0.1 ng/L I-TEQ/liter of wastewater, respectively. Recently, many alternative methods like microwave sterilization, alkaline hydrolysis, advanced steam sterilization, dry-heat sterilization have been suggested for the treatment of biomedical wastes.

The Bamako convention (African nations related), various environment and sustainable development conferences (Agenda 21), UN Committee of Experts on the transport of Dangerous goods, UN Economic Commission to Europe, and the Aarhus Convention are some of the other important treaties related to BMW management. The World Health Organization (WHO) publishes a policy paper on a regular basis titled "Safe healthcare waste management (WHO)" that consists of strategies and policies for countries related to the management of health waste [7, a regular basis titled "Safe healthcare waste management (WHO)" that consists of strategies and policies for countries related to the management of health waste [7, 8].

## National Policy

Every nation should develop a policy based on the resources and facilities available in the country, giving due credit to the cultural aspects of BMW handling. Adequate regulations with reference to collection, segregation, storage, handling, transport, and disposal should also be addressed in the policy. It also should include information on the responsibilities and training requirements for personnel involved in all sectors of healthcare as well as biomedical/ biotechnological researchers. The policy should contain the following, according to the WHO [1]: (i) BMW definition; (ii) Roles of personnel who handle and dispose of biomedical waste; (iii) Maintenance of record and reporting procedure (forms); (iv) Procedure for licensing and permit for treatment and handling of waste; (v) Precise laws for the management of biomedical waste; (vi) Rules for the protection of health and safety of workers and environment; and (vii) Special courts to handle issues due to non-compliance with law and accidents.

In India, the first biomedical waste management rules were framed in 1998 under the Ministry of Environment and Forest and the rules were regularly updated as per WHO guidelines [9]. The most recent update was proposed in 2016 under the Ministry of Environment, Forest and Climate Change, after WHO, in 2014, published updated guidelines on the safe management of waste from health care [1, 10]. Some of the key updates related to (a) Duties of the occupier (making them more responsible for biomedical waste management); (b) Duties of common biomedical waste treatment facility (CBMWTF)(duties are delineated better for record maintenance and reporting, barcoding and GPS for biomedical waste); (c) Accident reporting (forms are provided); (d) Waste management (effluent treatment of liquid waste); (e) Alternative advanced disposal methods (plasma pyrolysis, dry heat sterilization); (f) Impetus on segregation, packaging, transport and recycling of biomedical waste; are present in the guidelines [10].

## MANAGEMENT OF BIOMEDICAL WASTE

### Minimization of BMW

Management of BMW works on the concept of reducing, recycling and reusing (3 R's) with the main emphasis on minimization of production and minimal disposal of BMW. WHO proposes a waste hierarchy where different methods are classified based on desirability (*i.e.*) having an overall impact of the method on the environment, public health, financial affordability, and social acceptability. In order to minimize waste production, the following recommendations were made by WHO regarding [1] (a) source reduction (purchase of least possible starting material thereby lowering waste production, reducing chemical waste by replacing chemical treatment with physical treatment whenever possible, reduction of wastage of products); (b) hospital-level management measures (centralized purchasing of hazardous materials - minimize duplication and reduce wastage, monitoring of chemical usage); (c) stock management of hazardous chemical and pharmaceutical products; and (d) employees should be trained on waste minimization [1].

**Reuse, Recycling and Recovery of BMW**

Medical devices or materials can be categorized into three types: (a) disposable items (non-medical supplies, syringes); (b) medical devices that do not pose cross-infection threat (BP meters); and (c) reusable medical devices (surgical instruments). All these properties have to be considered before opting for reuse of medical items. Any combination of cleaning, disinfection, reconditioning, decontamination and sterilization steps is involved while reusing BMW [2, 3]. Sterilization methods recommended for reusable items are thermal sterilization (dry and wet sterilization), chemical sterilization (hydrogen peroxide, peracetic acid, hypo). The effectiveness could be checked by *Bacillus stearothermophilus* test and *Bacillus subtilis* test for thermal and chemical treatment, respectively.

Recycling is the next best option though reuse is the preferred choice since recycling involves high energy input to convert the waste into a product and also transportation cost (onsite to offsite). Recovery implies either direct recovery of energy in the form of fuel production from waste or indirectly composting of organic waste to produce compost or other agricultural products. Plastics (PET/PETE), glassware, cardboard, paper, wood and metals are some of the common non-hazardous materials encountered and as they are easily recyclable for energy production or degradability [5, 11, 12].

**Steps Involved in BMW Management**

Biomedical waste management involves segregation of waste, storage, transport of BMW, treatment (on-site or off-site), safe disposal. Since biomedical waste is highly diverse in nature (chemical/infectious agent/heavy metal/toxic/non-hazardous), a better understanding of the type of waste, nature of the hazard, the safety of personnel and the environment is required to devise a safe, effective and sustainable management system. Also, different sectors of the society are involved in the BMW management like hospital management, BMW handling personnel, laboratory cleaner/ maintenance people, storage and transportation people, common BMW facility, treatment & disposal facility, and hence a lot of coordinated effort is required for the safe disposal of BMW [13].

## *Segregation*

The risk posed on the environment and health by BMW is huge in a country like India, where many stray animals roam in different regions in an uncontrolled manner, and manual scavenging is still rampant. Some of the major risk factors include a) the spread of infectious disease; b) release of radioactive waste/ heavy metals like mercury from medical equipment; c) unused/ damaged/ expired antibiotics and d) other pharmaceutical products due to improper treatment of the various biomedical wastes. Since BMW are varied in nature, there is a need to segregate them into different categories for processing accordingly. The "three-bin system" is used commonly across the world to segregate the wastes into general/non-hazardous waste, infectious/ toxic waste and sharps waste [1]. As different categories of BMW should be stored separately and not mixed, the simplest way is to segregate them according to the source is by using colour-coded bags or containers (Table **1**).

**Table 1. Segregation of Biomedical Waste.**

| Type of Waste | Colour and Type of Container | Treatment or Disposal Options |
|---|---|---|
| Highly infectious (microbial cultures, swabs, excreta) | Yellow marked with "HIGHLY INFECTIOUS" - strong leak proof plastic or autoclavable container | Pretreated with non-chlorinated chemicals onsite followed by incineration/plasma pyrolysis |
| Other infectious waste, including anatomical and pathological waste | Yellow with biohazard label - leak proof plastic bag or container | Incineration/plasma pyrolysis/deep burial, in the absence of the above methods; autoclaving followed by shredding has to be done |
| Sharps waste | Yellow marked with SHARPS and biohazard label - puncture-proof container | Autoclaving/dry heat sterilization followed by shredding or mutilation or encapsulation and could be recycled for benches, hangers, bricks |
| Chemical and pharmaceutical waste | Brown with appropriate labels - plastic bag or rigid container | Incineration/plasma pyrolysis/encapsulation. In the case of drugs, they have to be returned to the manufacturer or supplier or CBMWTF for safe disposal |
| Radioactive waste | Labeled with radiation symbol - lead box | Waste is to be kept in an isolated and designated place until 10 half-lives have elapsed and then sent to an authorized centre or CBMWTF for safe disposal |
| Recyclable waste | Red - plastic bag or container | Autoclaving/microwaving/hydroclaving followed by shredding or mutilation and treated waste sent to recyclers for plastic or fuel or road making |

| Type of Waste | Colour and Type of Container | Treatment or Disposal Options |
|---|---|---|
| Glasswares, metallic implants | Blue cardboard boxes | Autoclaving/hydroclaving/microwaving and sent for recycling |
| General health care waste | Black plastic bag, flowers, food waste, wood, cardboard | Could be segregated further to recyclable, biodegradable and non-recyclable and disposed off accordingly. Some wastes could be recycled as compost or animal feed or energy production |

## Storage of Biomedical Waste

The Guidelines for the design and construction of hospitals and health care facilities have outlined the basic guidelines and specifications for the BMW storage facility. Since the biomedical waste needs to be collected in sufficient quantities before it could be treated or transported to CBMWTF, storage becomes a necessity in some cases. The storage facility should be located inside the healthcare facility and should be a dedicated site. The storage facility should be far off from the food preparation area and should have various signs like no eating/drinking/smoking, restricted entry, biohazard/ radiation/toxic chemical. Various symbols are universally used to denote the different hazardous nature of biomedical wastes; the examples include corrosive, inflammable, toxic, irritant, and explosive [1].

Four types of waste storage facilities based on the type of waste include

i. general/non-hazardous waste,
ii. hazardous waste,
iii. infectious and sharp waste,
iv. chemical and pharmaceutical waste.

The biomedical waste is varied in nature and hence different types of containers are routinely employed in different countries. The waste container could be of varying sizes and shapes depending upon the use. The containers should be made up of sturdy material and should be leak-proof, puncture-proof, impermeable, and color-marked, and should have a tight-fitting removable lid. The plastic should be of at least 70 μm thickness and autoclavable. The waste bags and containers should be filled to less than 3/4$^{th}$ of their capacity and should be sealed or tied but never stapled to avoid leakages. The wastes should be labeled immediately, and the label should have the following details - date of collection, type of sample/waste, source of waste to facilitate proper and effective disposal.

According to the WHO, the storage facility should be impermeable, have a solid floor with a good drainage facility, exhaust and water facility. The facility should be cleaned regularly, accessible to transport vehicles, have cleaning/spillage containment equipment, and should have protection against animals, birds, insects, and unauthorized people. Proper records of arrival, treatment, and disposal dates for each biomedical waste have to be maintained. Similarly, documentation of spillages, accidents, cleaning, inspection, and repair/maintenance work should also be maintained properly. The MSDS (material safety data sheet) should be available along with instructions on precautions and treatment procedures should also be available in case of chemical or toxic wastes. The personnel involved in storage and transport should be trained on a regular basis on the current trends in waste management and safety measures for BMW.

## Transport of BMW

The hazardous and non-hazardous waste should be transported separately and not together. Three transport systems have been proposed: (i) general or non-hazardous waste in black colored containers, (ii) infectious and sharp waste in yellow containers/ bags and (iii) other infectious waste like chemicals, drugs. Dedicated trolleys should be used for the different wastes generated and should possess the following qualities, *i.e.* should not have sharp edges or damaged, easy to clean, pull/push and load/unload, lockable and appropriate size and shape. Transporting BMW to offsite treatment requires the transporting company and other people involved in complying with national policy on BMW management or international policy like the United Nations, WHO guidelines [18]. The vehicle should be of appropriate size and design as per international convention with a proper securing mechanism to prevent spillage. The drivers should be medically fit and also trained in basic aspects like type of BMW, relevant legal regulations, safe handling of wastes, safety measures & contact information in case of accidents, documentation. The vehicle should have proper labels like a biohazard, toxic substance, flammable, infectious substance, *etc.* as per international conventions and type of BMW. The driver should carry a consignment note carrying information such as (a) class, other details of waste and its source, (b) pick up date, (c) destination, (d) driver name, (e) contact details (source and destination), (f) authorized person's signature (source and destination) [6, 14 - 18].

## Treatment and Disposal of BMW

The treatment of biomedical waste is meant to reduce, minimize or neutralize the hazards posed by the biomedical waste [1, 3, 13, 19 - 22]. The following factors have to be considered before setting up of BMW treatment facility - (a) type and characteristics of biomedical waste, (b) availability of treatment method, (c) economics of treatment facility, (d) capacity and efficiency of the treatment method, (e) infrastructure and maintenance of the treatment facility, (f) skills and training of the personnel handling the facility, and (g) environmental and public safety [23, 24]. The common treatment methods are briefly discussed below.

## Thermal Processes

Thermal processes involve heat either in the form of wet heat or dry heat and can be classified into high-heat systems and low-heat systems. In high heat systems, the BMW are subjected to very high heat that could lead to combustion of organic matter and different atmospheric emissions. An example of high-heat system is pyrolysis, where wastes are subjected to high heat in the absence of oxygen, and this process is carried out for organic substances. In case of low-heat systems, the temperature ranges from 100-180 °C and heat involved is either dry heat or moist heat (steam); examples include autoclaving, microwave treatment (moist heat) and thermal radiation/ conduction or convection using IR heaters (dry heat).

## Steam Based Technologies

### Autoclaves

Autoclaves are simple devices that have been routinely used for many years for treatment of wastes by laboratories and hospitals. These processes involve the utilization of steam (wet heat) to kill pathogens and the key factors that affect the process are time, pressure, temperature, penetration of heat, load size, stacking of bins, package material, *etc.* Hence simple materials like microbial cultures, liquids (blood, body fluids), used sharps, gauze material, cotton, glassware, plasticware and small instruments could be autoclaved. The materials that should not be autoclaved include solvents (acids, alcohols), radioactive wastes, chemical waste, toxic wastes, semi-organic compounds, heavy metals, anatomical wastes, large and bulk materials.

The basic steps involved in the autoclave process are (i) waste collection, (ii) pre-heating, (iii) waste loading, (iv) air evacuation, (v) steam treatment, (vi) steam discharge, (vii) unloading, (viii) documentation and (ix) mechanical treatment of wastes (if required). A classical autoclave consists of a steel chamber that can withstand high temperatures and pressures for long periods of time, surrounded by a jacket with water lines (inlet and outlet) enabling entry and exit of steam (into chamber and jacket), gauges (pressure and temperature) and timer. The quality air evacuation determines the success of sterilization of the wastes as air provides very good insulation, thereby preventing heat penetration. The air released from waste treatment autoclaves should be filtered through high efficiency particulate air (HEPA) filters to prevent pathogenic aerosols from entering the environment. Based on air removal processed, autoclaves are classified into gravity-displacement autoclaves, pressure-pulse autoclaves, and pre/high vacuum autoclaves. Among these, pre-vacuum autoclaves are routinely used in labs due to their greater efficiency and cost-effectiveness. The prescribed conditions for sterilization of wastes by autoclaving are temperature 121 °C that corresponds to 15 psig or 30 psia (1-2 bar gauge pressure) for about 15-30 minutes. For additional controls, colour changing strips (containing thermochromic agents) or integrators or Bowie-Dick test packs could be used to test the efficiency of the autoclaving method [25].

## *Integrated Steam Based Treatment Systems*

These are second-generation steam-based systems that combine other mechanical processes with steam treatment to improve penetration of heat leading to uniform distribution of heat, thus resulting in greater sterilization. Other advantages include breaking of waste, and as different methods are combined in these systems, the performance could be continuous rather than in batches. These systems are also called advanced autoclaves, hybrid autoclaves, or advanced steam treatment technologies. The mechanical processes could be combined with steam treatment either before or after or during steam sterilization and hence according to the WHO (2014) [1], the following examples are possible; (a) steam treatment–mixing- fragmenting, followed by drying and shredding; (b) internal shredding followed by steam treatment – mixing and then drying; (c) internal shredding – steam treatment-mixing followed by drying; (d) internal shredding followed by steam treatment – mixing-compaction.

## Microwave Based Technologies

This technology works on the principle that when wastes are subjected to a frequency of about 2450 MHz and a wavelength of 12.24 cm, the water in the wastes gets heated and converts into steam. The sterilization occurs due to moist heat and steam generated by microwave energy. The wastes could be treated batch-wise or in a semi-continuous manner; a typical system consists of a chamber and 2-6 megatons (microwave generator) capable of producing 1.2 kW output. One typical cycle would range from 30-60 minutes and depending upon the type and quantum of waste, multiple cycles are performed. The various components of a semi-continuous microwave system comprise of the automatic charging system, hopper, shredder, conveyor screw, steam generator, microwave generators, discharge screw, secondary shredder, and controls (microprocessor-based HEPA filters to prevent airborne pathogens from entering the environment). The types of wastes treated by microwave-based technologies are similar to that of autoclaving processes.

## Dry Heat Technologies

Dry heat has been used to sterilize glassware and plasticware for many years in laboratories using hot air ovens. The wastes are treated with dry heat and they get heated by convection (forced - air heated by resistance heaters/natural - through walls of the hot chamber), conduction, or thermal radiation (IR or quartz heaters). These methods have limited utility as the exposure time and temperatures are higher than steam-based technologies. The spores of the organism *Bacillus atrophaeus* are used as a control as they are resistant to dry heat sterilization.

## Chemical Processes

Chemical processes involve the interaction of key functional groups/molecules present in the pathogens or wastes with the treated chemicals resulting in neutralization or degradation of the molecules, thereby killing or inactivating microorganisms. The common chemicals used are disinfectants like chlorine dioxide, lime, phenolic compounds, sodium hypochlorite (bleach), peracetic acids, ozone gas or calcium oxide, ammonium salts (inorganic). To maximize the interaction between the chemical and waste, the waste could be shredded, ground and mixed together with the chemical. In the case of liquid waste, the treated mixture has to be processed to remove water to remove/recycle the chemicals. Alternatively, heated alkali could also be used for digesting pathological wastes (anatomical parts - humans and animals, tissues).

The major factors that affect chemical treatment are the type and amount of chemical, the type and organic content of waste, contact time between waste and chemical, and environmental factors like pH, temperature. The major drawbacks are some of the chemicals/disinfectants that can cause irritation to skin/eye, mucous membranes, being corrosive, toxic and hazardous in nature. Alkali hydrolysis is a method routinely used for the treatment of anatomical wastes, where the tissues are soaked in alkali in a proportional manner and heated to 110 - 127 °C or higher for 6-8 hours for complete digestion. The alkali treatment could also neutralize or degrade hazardous chemicals like aldehydes, therapeutic agents.

## Irradiation Processes

Electron beams or UV irradiation or electromagnetic radiation (EMR) from Co-60 could be used to kill pathogens present in the biomedical wastes. The facility should have a protective mechanism to protect the operators from getting exposed to EMR. The success of these methods depends on the dose and tissue penetration of the radiation and hence could be added as a supplemental process to primary treatments like thermal processes. For example, germicidal UV radiation could kill airborne microorganisms but could not penetrate closed plastic bags.

## Biological Processes

The processes that utilize either enzymes or living organisms to degrade the BMW belong to biological processes. Examples include landfill methods, composting and vermiculture that have been successfully utilized to degrade BMW, especially organic wastes. The major issue could be the spread of disease to livestock as well as humans if the BMW was not processed completely and sent for disposal. Landfills are the last disposal method for treated BMW and two types of landfills are possible - (a) controlled landfilling (preferred method) as it prevents easy access to stray animals and scavengers and protects the environment and (b) uncontrolled dumping (not preferred). The most preferred choice of landfills includes sanitary landfill (highest priority), engineered landfill (medium priority) and controlled landfill (low priority). These landfills allow to prevent soil pollution, water (surface and groundwater) pollution, air pollution, block/prevent access to pests, vectors, animals and public.

## Mechanical Processes

Mechanical processes involve breaking, shredding, grinding, mixing and compaction of the BMW, primarily meant for used sharps like needles, broken scalpels, *etc*. These processes do not kill pathogens and hence are used in combination with other treatment methods to maximize the efficiency of the treatments by increasing the exposure or surface area of waste to treatment.

The major drawback is that aerosols are generated during mechanical processes and when untreated or live pathogens are present in the BMW, it could result in the spread of airborne pathogens. Hence mechanical processes are generally used as part of a closed treatment system so that the air coming in and going out is disinfected properly before release. Also, prior treatment or disinfection of the sharps is preferred before breaking the sharps using blenders or shredders.

## Incineration, Pyrolysis and Gasification

All the three methods use high temperatures to convert organic waste (combustible) to inorganic matter (incombustible), thereby leading to reduction in a significant volume of waste. The type of products (gas, liquid and solid) in the three methods also varies due to the differences in the treatment methodology. The difference in treatment is in terms of stoichiometric ratio, type of atmosphere used for heating, reaction temperatures. Many types of incinerators are available like starved-air incinerator, multiple-chamber incinerator, rotary kilns, small-scale incinerators, co-incineration plants. Since pollutants belonging to all three phases (gas, liquid and solid) are produced by these methods, stringent policies are needed to regulate and protect both living organisms and the environment. These methods are used to treat anatomical wastes, carcasses, chemical wastes and pharmaceutical waste.

## Encapsulation and Inertization

Encapsulation is a process where the biomedical waste is filled in a high-density container (metallic or plastic) up to 3/4$^{th}$ capacity, followed by the addition of immobilizing materials (plastic foam, bituminous sand, cement or clay) and sealing the container. The pharmaceutical waste, sharps or chemical wastes are encapsulated and disposed into landfills. This process prevents access of these wastes to scavengers, thereby protecting them from injury and infection. Inertization is another method that is employed to neutralize toxic waste by treating the waste with lime and cement. A typical inertization mixture, according to WHO (2014), would have 65% pharmaceutical waste, 15% lime, 15% cement and 5% water [1]. All the ingredients are mixed together and then allowed to solidify into small cubes of $1m^2$ or transported to previously landfill municipal waste and deposited to solidify. Inertization and encapsulation are used as the last alternative treatment methods when other methods are not available.

## Emerging Technologies

New technologies, so far discussed in the literature, utilized for treatment of biomedical wastes are plasma pyrolysis, promession, ozone, superheated steam, and nanotechnology. Since many of these are relatively new techniques, they do not have any track record for their use in biomedical waste treatment. A lot of work is required to validate their application in BMW management before being adopted by hospitals and labs for routine use. In the superheated steam method, the temperatures of 500 °C are obtained using superheated steam and used to kill pathogens and neutralize toxic wastes. The vapors obtained are then heated to 1500 °C in a steam reforming chamber. In the case of plasma pyrolysis, ionized gas in the plasma state is used to pyrolyze the wastes. When the gas in plasma is subjected to plasma torches or electrodes in an atmosphere with little or no air, it results in the generation of very high temperatures that could be used for sterilization of waste. Both of these methods are very expensive and also need sophisticated materials to prevent pollutants from entering the environment.

Ozone is another promising method for killing bacteria since it gets converted easily to oxygen post-treatment. This method needs a mixer or shredder to break and mix the wastes with ozone for proper treatment, and is used for water and air treatment. Controls are necessary to ensure that bacteria are neutralized with ozone treatment prior to the release of waste for other treatments. Promession is an emerging method to treat anatomical wastes and uses mechanical processes and freezing technology. Liquid nitrogen is used to freeze the samples and mechanical vibration is used to break the anatomical waste into powder before burial. It allows the recovery of metal parts from anatomical waste and also decomposes fast due to reduced mass and volume. Silver nanoparticles have been shown to have bactericidal properties and could be used as disinfectants though concerns regarding bacteria developing resistance to silver need to be addressed [26, 27]. Other emerging technologies include supercritical water oxidation, vitrification, gas-phase chemical reduction, Fe-TAML/peroxide treatment, mechanochemical treatment, electrochemical technologies, biodegradation, and phytotechnology [1 - 3].

## CONSENT FOR PUBLICATION

Not applicable.

## CONFLICT OF INTEREST

The authors confirm that this chapter contents have no conflict of interest.

## ACKNOWLEDGEMENTS

Declared none.

## REFERENCES

[1]     World Health Organization. Safe Management of Wastes from Health-Care Activities. World Health Organization 2014.

[2]     Capoor MR, Bhowmik KT. Current perspectives on biomedical waste management: Rules, conventions and treatment technologies. Indian J Med Microbiol 2017; 35(2): 157-64.
        [PMID: 28681801]

[3]     Ali M, Wang W, Chaudhry N, Geng Y. Hospital waste management in developing countries: A mini review. Waste Manag Res 2017; 35(6): 581-92.
        [http://dx.doi.org/10.1177/0734242X17691344] [PMID: 28566033]

[4]     UNEP (United Nations Environment Programme). Basel convention on the control of transboundary movements of hazardous wastes and their disposal. Geneva, Secretariat of the Basel Convention. 1992.

[5]     Secretariat of the Basel Convention, United Nations Environment Programme. Technical guidelines on the environmentally sound management of biomedical and healthcare wastes: (Y1; Y3). Secrt Basel Convention.

[6]     No. 28911. Basel convention on the control of transboundary movements of hazardous wastes and their disposal. Concluded at Basel on 22 March 1989. United Nations Treaty Series. 1999. 415.

[7]     Secretariat of the Basel Convention, World Health Organization. Preparation of National Health-care Waste Management Plans in Sub-Saharan Countries: Guidance Manual 2005.

[8]     Secretariat of the Basel Convention. Technical guidelines on environmentally sound management of biomedical and health care waste 1992.

[9]    The Gazette of India Biomedical Wastes (Management and Handling) Rules, India: Ministry of Environment and Forests, Government of India Notification Dated 20th July 1998.

[10]   Bio Medical Waste Management Rules Published in the Gazette of India, Extraordinary, Part II, Section 3, Sub Section (i), Government of India Ministry of Environment, Forest and Climate Change Notification; New Delhi, the 28th March 2016.

[11]   Rushbrook P, Zghondi R. Better health care waste management: an integral component of health investment. World Health Organization 2005.

[12]   Sharma P, Sharma A, Jasuja ND, Somani PS. A Review on biomedical waste and its management. Sign Bioeng Biosci 2018; 1(5).
[http://dx.doi.org/10.31031/sbb.2018.01.000522]

[13]   Datta P, Mohi GK, Chander J. Biomedical waste management in India: Critical appraisal. J Lab Physicians 2018; 10(1): 6-14.
[http://dx.doi.org/10.4103/JLP.JLP_89_17] [PMID: 29403196]

[14]   UN. Recommendations on the transport of dangerous goods. In: Recommendations on the Transport of Dangerous Goods: Model Regulations. New York: UN 2010; pp. 1-8.
[http://dx.doi.org/10.18356/da5b8427-en]

[15]   UN. European Agreement Concerning the International Carriage of Dangerous Goods by Road (ADR): Applicable as from 1 January 2010. New York: UN 2011; 1297 pages.
[http://dx.doi.org/10.18356/fd72a1d4-en]

[16]   UN Recommendations on the Transport of Dangerous Goods - Model Regulations. Sixteenth revised edition. New York: UN 2009. http://www.unece.org/trans/danger/publi/unrec/rev16/16files_e.html

[17]   Secretariat of the Basel Convention on the Control of Transboundary Movements of Hazardous Wastes and Their Disposal. Technical Guidelines on the Environmentally Sound Management of Biomedical and Healthcare Wastes (Y1,Y3). Geneva: UN 2002. p. 72.

[18]   International Atomic Energy Agency.. Application of the Concepts of Exclusion: Exemption and Clearance: Safety Guide. IAEA Safety Standards Series RS-G-17 IAEA Safety Standards Series RS-G-17 2004.

[19]   World Health Organisation Staff. Guidelines for Safe Disposal of Unwanted Pharmaceuticals in and After Emergencies. 1999.

[20]   U.S. Government. Reference Guide to Non-Combustion Technologies for Remediation of Persistent Organic Pollutants in Stockpiles and Soil. Books LLC 2011.

[21]   Emmanuel J. Non-incineration medical waste treatment technologies. Washington, DC: Health Care Without Harm 2001.

[22]   Emmanuel J, Stringer R. For proper disposal: A global inventory of alternative medical waste treatment technologies. Arlington: Health Care Without Harm 2007.

[23]   WHO (World Health Organization). Review of health impacts from microbiological hazards in health-care wastes. Geneva: World Health Organization 2004.

[24]   WHO (World Health Organization). Survey of hospital wastes management in South-East Asia Region. New Delhi: World Health Organization 1995.

[25]   Lemieux P, Sieber R, Osborne A, Woodard A. Destruction of spores on building decontamination residue in a commercial autoclave. Appl Environ Microbiol 2006; 72(12): 7687-93.
[http://dx.doi.org/10.1128/AEM.02563-05] [PMID: 17012597]

[26]   Chopra I. The increasing use of silver-based products as antimicrobial agents: a useful development or a cause for concern? J Antimicrob Chemother 2007; 59(4): 587-90.
[http://dx.doi.org/10.1093/jac/dkm006] [PMID: 17307768]

[27]   Miller G, Senjen R. Nanotechnology in Food and Agriculture. Nano Meets Macro 2010; pp. 417-44.

# CHAPTER 13

# Microbial Pollution Control through Biogenically Synthesized Silver Nanoparticles (*Bacillus Spp*)

**J. Senthil Kumar[1,*], R. Sathya[2], P. Balamurugan[3], R. Gomathi[1], B. Preetham Kumar[1] and R. Vishnu[1]**

*[1] Department of Biotechnology, Sri Krishna Arts and Science College, Coimbatore, Tamil Nadu, India*

*[2] Department of Information Technology, Sri Krishna Adithya College of Arts and Science, Coimbatore, India*

*[3] Government Arts College, Coimbatore, Tamil Nadu, India*

**Abstract:** Microbes are omnipresent in nature, and have both beneficial and non-beneficial activities for humans. In the system we live in is getting worse, due to microbes, that obviously end up in pandemic diseases like Covid-19. This chapter made an attempt to control pathogenic microorganisms either in the form of a spray or incorporated in disinfecting agents. Nanotechnology deals with the synthesis, characterization, and manipulation of metals at the nanoscale. The nanoparticles are precisely used due to their smaller size, physical properties, *etc.* which have shown a change in other materials that are in contact with these tiny particles. They are synthesized through various methods such as physical, chemical, and biological means. This study is aimed at the use of biological, eco-friendly, and green synthetic nanoparticles due to their less time consumption and ease. The visual observation was made with the color change indicated by the synthesized nanoparticles. They were further characterized by UV-Visible Spectrophotometer, XRD, EDAX, Zeta analysis, and FESEM. The size of nanoparticles was about $5.49 \pm 2.10$ nm. The synthesized nanoparticles showed significant results in control of *S. aureus* and *P. aeruginosa* with the zone of inhibition having a size of 16 mm and 18 mm, respectively. The green synthesized nanoparticles were based on the minimal inhibitory concentration and minimal bactericidal concentration.

**Keywords:** *Bacillus Safensis*, BSNPs, Covid-19, Cytotoxicity and Well Diffusion Assay, EDAX, MIC, MBC, SEM, UV-VIS Spectrophotometer.

## INTRODUCTION

The members of genus *Bacillus* are rod-shaped, gram-positive, gram variable,

---

\* Corresponding author J. Senthil Kumar: Department of Biotechnology, Sri Krishna Arts and Science College, Coimbatore, Tamil Nadu, India; Tel: 9486636577; E-mail: senthil.btjeyaraj@gmail.com

**J. Senthil Kumar, P. Ponmurugan & A. Vinothkanna (Eds.)**

spore-forming, aerobic, anaerobic, and ubiquitous bacteria, primary habitat of soil, and sometimes in water as well. The term *Bacillus* may literally be defined as cylindrical or rod-like bacteria, often occurring in chains. They are omnipresent environmental organisms, frequently isolated in laboratory as contaminants of media and specimens. It is known to associate with the production of enzymes, antibiotics and potential secondary metabolites with an array of applications in the field of medicine, agriculture and pharmaceuticals, *etc*. The antibiotic Bacitracin was determined to be effective on gram-positive bacteria only [2], *B. polymyxa* produces Polymyxin B [8]. *B. amyloliquefaciens*, often associated with certain plants that are capable of synthesizing different antibiotics like of Bacillaene, Macrolactin and Difficidin [12]. Sources of *Bacillus Spp* in foodborne outbreaks have been described including rice, meatloaf, turkey loaf, mashed potatoes, beef stew, apples and hot chocolate sold in vending machines [7]. Even though, it is not recognized as a major human pathogen, it is responsible for immunosuppressed patients and being opportunistic pathogens. Cells were mesophilic, aerobic, chemoheterotrophic, Gram-positive, spore-forming rods that are motile by means of polar flagella. Cells ranged in 0.5 – 0.7 μm in diameter and 1.0 – 1.2 μm in length. Growth occurs at 0 – 10% (w/v) NaCl at pH 5.6. Growth occurs at 10–50 °C (optimum, 30 – 37 °C) but not at 4 or 55 °C. Colonies are round, undulate, dull white, non-luminescent and have irregular margins on TSA plates incubated at 32 °C for 24 hours.

## NOSOCOMIAL PATHOGENS

*Klebsiella pneumonia,* a Gram-negative bacteria that can cause different types of healthcare-associated infection such as pneumonia, bloodstream infections, wound or surgical site infection and meningitis. Healthy people usually do not get *Klebsiella* infections. *Pseudomonas aeuroginasa,* a bacteria tend to live and breed in water, soil and damp areas. The warmer and wetter, the better conditions for the bacteria to multiply. People who are in the hospital for surgery or treatment for a major illness are most vulnerable to this kind of infection [13]. Carbapenems (*e.g*, Imipenem, Meropenem) with anti pseudomonal quinolones may be used in conjunction with an aminoglycoside [3].

## SYNTHESIS AND CHARACTERIZATION OF BSNPS

Silver NPs are of interest due to its unique properties which can be incorporated into antibacterial applications, biosensor materials, composite fibers, cryogenic super-conducting materials, cosmetic products and electronic components. The silver nanoparticles can be synthesized by means of physical, chemical and green or biological synthesis approaches it had further characterized. Bacteria have also been explored in the synthesis of silver NPs and reported that highly stable silver nano particles could be synthesized by bioreduction of aqueous silver ions with culture supernatant of non-pathogenic bacterium [5]. The morphology, size, and shape of the silver nanoparticles were examined using field emission electron microscopy and particle size analysis [1].

## WELL DIFFUSION ASSAY

The Minimum inhibitory concentration (MIC) and minimum bactericidal concentration (MBC) of synthesized nanoparticle state the lowest concentration that will inhibit a visible growth of a microorganism after overnight incubation. The antimicrobial activity of the synthesized nanoparticle was performed by well diffusion assay [6].

## MATERIALS AND METHODS

### Isolation and Collection of *Bacillus safensis*

*Bacillus safensis* was isolated by Narendhran. S, *et al.*, 2019, during their screening of soil samples against commonly encountered soil pathogens. The 16S rRNA gene sequence of the strain NS22 was indicated as a novel strain with 82% and the same was submitted in the NCBI database with accession Number MK168569. The *Bacillus safenisis* N22 was used in this study.

### Inoculum Preparation

The inoculum was prepared by inoculating in 5 mL nutrient broth with 18 hours old culture of *Bacillus safensis*, which was incubated at 37°C for 16 to 24 hours. Similarly, slant cultures were also prepared and revived sequentially to ensure the viability of the bacterial cells.

### Synthesis of Biogenic Silver Nanoparticles

The starter culture of *B. safensis* was incubated overnight at 37°C for 48 hours in nutrient broth. After incubation, the culture was centrifuged at 10,000 rpm for 30 minutes and the supernatant was filtered through Whatman No.1 filter paper. The culture filtrate was then added to an equal amount of 0.1 M silver nitrate solution. The synthesis was further carried out under dark conditions at 35°C for 48 to 72 hours [10].

The biogenically synthesized silver nanoparticles were isolated by centrifugation at 10,000 rpm for 20 minutes. The pelleted silver nanoparticles were washed twice with sterile distilled water by centrifugation at 10,000 rpm for 20 minutes to remove the excess silver ions and the culture filtrate. The isolated silver nanoparticles were freeze-dried for 12 hours and, thus, silver nanoparticles were obtained as powder, further grinded in a sterilized mortar and pestle under the sterile condition at the laminar air flow (LAF), and was used for characterization and antimicrobial studies.

## CHARACTERIZATION OF BIOGENIC AGNPS

### Visual Characterization

Visual characterization was observed contrastly as the supernatant taken in one conical flask, similarly the supernatant and silver nitrate in another conical flask were kept for incubation at 24-48 hrs at 37°C, the results were noted.

### UV-VIS Spectroscopy

UV-VIS Spectroscopy (SYSTRONICS, 2202), the reduced silver nanoparticles were further characterized by UV-VIS spectroscopy at the range of 200-800nm. The results were noted and interpreted.

### FESEM Analysis

SEM analysis (MALVERN, v2.1), size and shape of the silver nanoparticles was determined by 3 dimensional scanning electron microscope and particle size. The results were noted.

### EDAX Analysis

EDAX analysis (BRUKER), the presence of elemental silver was determined through EDX analysis, other compounds if any were also noted.

### Well Diffusion Assay

The biogenically synthesized silver nanoparticles were tested for further screening against selected bacterial pathogens [15]. Each inoculum was prepared by inoculating 5 mL of nutrient broth with 18 hours old culture of *K. pneumoniae, P. aeuroginosa, E. feacalis* and *S. aureus*, which were incubated at 37°C for 16 to 24 hours.

The sterile cotton swab was dipped into the broth and drained carefully without contaminating it and the lawn culture of bacteria was spread onto nutrient agar plates in triplicates. Only five wells of not closer than 24 mm from the center to center were made with sterile cork borer and synthesized silver nanoparticles as 20 and 40 μg/mL were evaluated, similarly DMSo as negative, streptomycin 20 μg/ mL as a positive control and the culture supernatant were also added in respective wells, kept for incubation at 37°C for 24 hours. The results were interpreted using the measurement of zone of inhibition in millimeter. The test was done in triplicates to ensure better results. The average results of the zone of inhibition were arrived following the standard deviation analysis.

## RESULTS

### Isolation of Bacterial Strain

The isolated bacterial colonies as *Bacillus safensis* SN22 were characterized and the molecular identification tool was $16_S$ RNA sequencing. The sequence was submitted in the NCBI gene bank with accession number MK168569. The phylogenetic mapping revealed 82% similarity index. The bacteria were sub-cultured and used throughout the study.

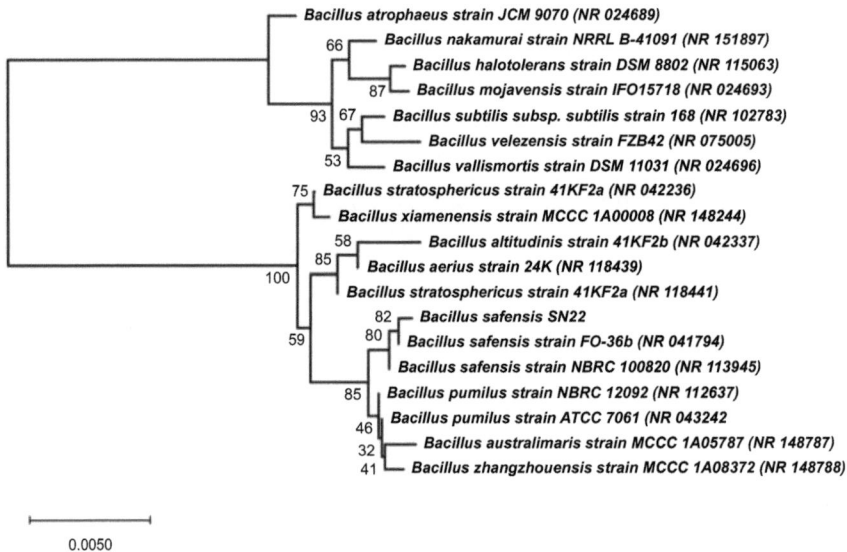

**Fig. (1).** Phylogenetic Map of *Bacillus safensis* SN22.

## BIOGENIC SYNTHESIS OF SILVER NANOPARTICLES

Upon addition of silver ions into the bacterial culture supernatant, the sample changed its colour from almost pale yellow colour to brown, with intensity increasing during the period of incubation as visualized in Fig. (**2**).

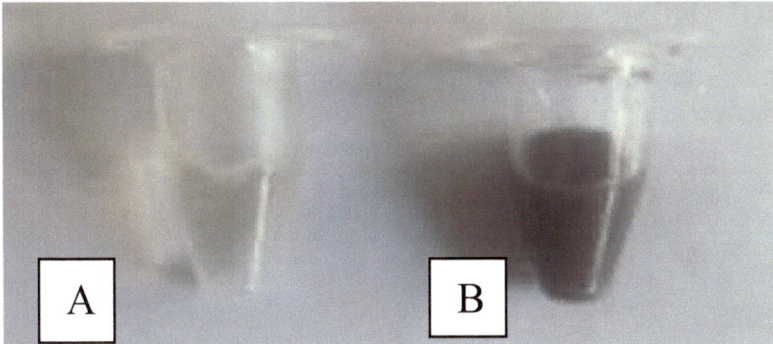

**Fig. (2). A** - culture supernatant with silver nitrate (pale yellow colour) **B** - synthesized silver nano particles (brown colour).

## CHARACTERIZATION OF BIOGENIC SILVER NANOPARTICLES

## Visual Characterization

The visual appearance of colour change was observed within 48 to 72 hrs, the colour change to brown precipitate indicated the presence of synthesized silver nanoparticles. Both, preliminary and visual characterizations were carried out to confirm whether the silver nitrate was reduced into silver nanoparticles mediated by biological method. Whether to go for further analysis or characterization of BSNPs would be an initial step. This may avoid wastage of time and expenses insured throughout the research (Fig. **3**).

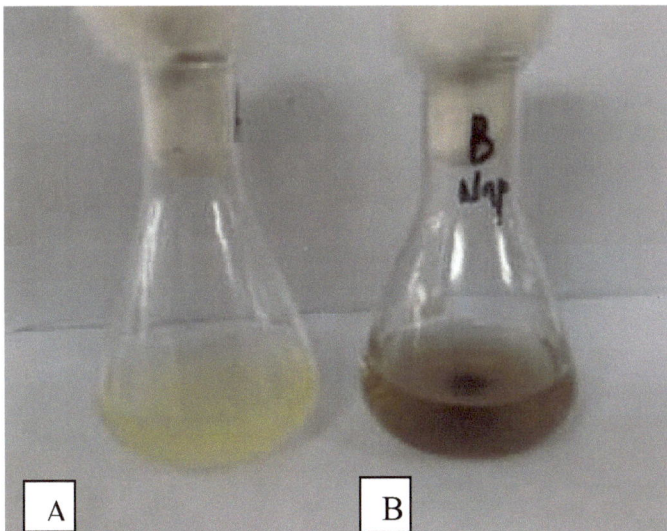

**Fig. (3). A**-(control)-culture supernatant; **B**-(test)-Biogenically synthesized silver nanoparticles.

## ULTRAVIOLENT SPECTROPHOTOMETRY

The absorption spectrum was recorded using Shimadzu UV 2400 PC series for the synthesized silver nanoparticles in the range of 200-800nm. The spectrum showed the absorbance peak at 430 nm corresponding to the characteristic surface plasmon resonance band of silver nanoparticles. The overall observations suggest that the bioreduction of silver ions was confirmed by UV-Visible spectroscopy (Fig. **4**).

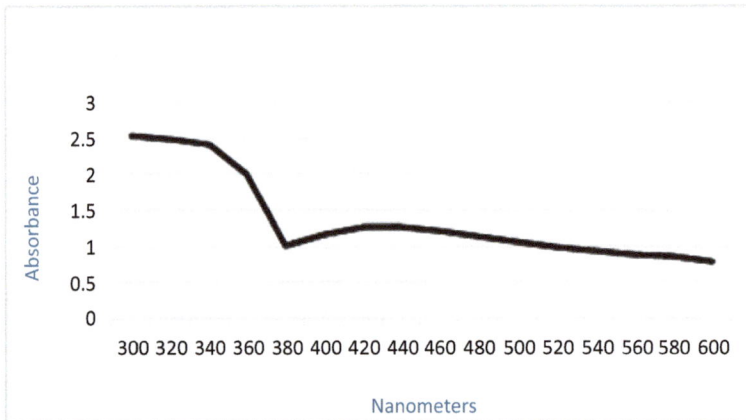

**Fig. (4).** UV-Visible spectral analysis of biogenic synthesized nanoparticles.

## EDAX AND SEM ANALYSIS

The EDAX analysis and SEM image revealed the elemental composition profile of the synthesized nanoparticles, which suggests silver as the constituent element using the FEG QUANTA 250 series (Fig. **5**).

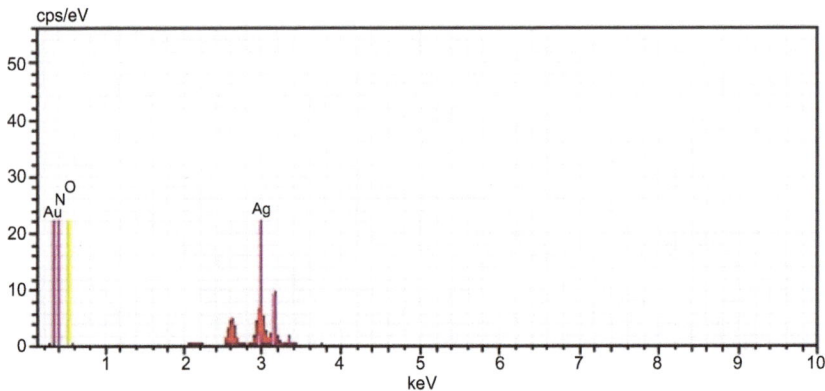

**Fig. (5).** EDAX analysis of biogenically synthesized nanoparticles.

**Table 1. Results with zone of inhibition in mm.**

| Wells | *Klebsiella pneumoniae* | *Pseudomonas aureginosa* | *Enterococcus faecalis* | *Staphylococcus aureus* |
|---|---|---|---|---|
| Positive control 20 µg/ml (streptomycin) | 21mm ± 0.9mm | 28mm ± 1.1mm | 35mm ± 0.8mm | 26mm ± 1.5mm |
| Negative control | - | - | - | - |
| Supernatant 20µl | - | - | - | - |
| BSNPs 20 µg/ml | - | 16mm ± 1.3mm | 14mm ± 1.5mm | 11mm ± 1.6mm |
| BSNPs 40 µg/ml | 8mm ± 0.8mm | 18mm ± 1.5mm | 17mm ± 1.9mm | 16mm ± 1.3mm |

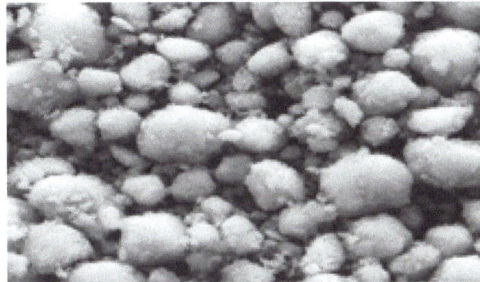

**Fig. (6).** FESEM analysis of biogenically synthesized nanoparticles.

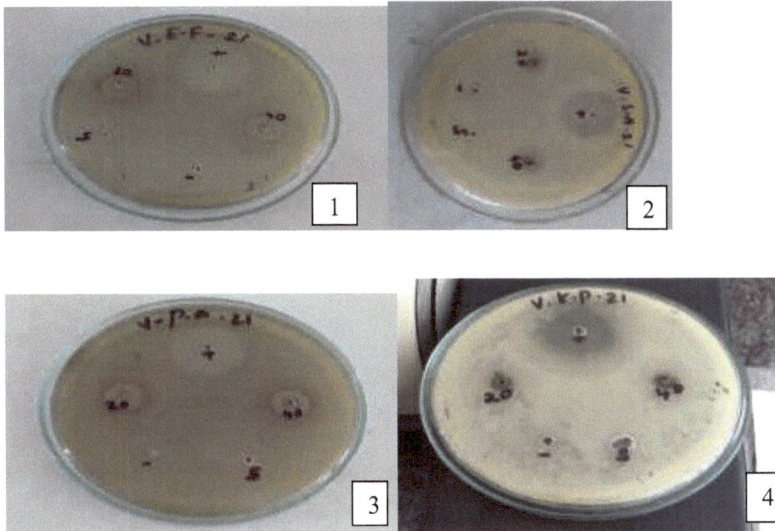

**Fig. (7).** Well diffusion assay against *K. pneumoniae, P. aeuroginosa, S. aureus* and *E. feacalis. 1 – E. faecalis*, 2 – *S. aureus*, 3 - *P. aeuroginosa*, 4 -*K. pneumoniae.*

The optical absorption peak near 3 keV indicates the presence of nano-sized metallic silver. However, other elemental signals, such as N and O, were also recorded, which are possibly due to emissions from proteins or enzymes present in the culture supernatant.

The characterization of synthesized silver nanoparticles was done by the FESEM analysis. The morphology of silver nanoparticles was predominately spherical and aggregated into larger irregular structures with well-defined morphology observed in micrograph (Fig. **6**).

## WELL DIFFUSION ASSAY

This test also suggests that using repeated doses of antibiotics makes bacteria resistant to commonly used antibiotics. In order to avoid this problem, biogenically synthesized silver nanoparticles at 20 and 40 µg/ml were used in the control of specific pathogens. The tests were carried out in triplicates to arrive at concordant values. Streptomycin had shown a greater zone of inhibition; the result with 40 µg/ml indicated positive results for *Pseudomonas, Enterococcus,* and *Staphylococcus*. Similarly, 40 µg/ml had shown positive control of pathogens as visualized.

## DISCUSSION

A detailed study on the biogenesis of silver nanoparticles by *B. safensis* was carried out, and its antibacterial effect on nosocomial pathogens was reported in this work. The biogenically synthesized silver nanoparticles were characterized for their shape, size, absorbance spectral value, and functional groups. The brown color is attributed to the excitation of SPR, as shown in Fig. (**4**), a characteristic and well-defined SPR band for silver nanoparticles is obtained at around 430 nm [4]. The image obtained by the FESEM also revealed the presence of spherical nanoparticles (Fig. **6**), confirming the result obtained by FESEM. Generally, metallic silver nanocrystals show a typical optical absorption peak approximately at 3 keV due to their surface plasmon resonance [9]. The growth inhibition effect of Bio-Ag0-6 on the Gram-negative strains *E. coli*, and *P. aeruginosa* and the Gram-positive strain, *S. aureus,* was quantitatively determined by the MIC and MBC methods [11]. 20 and 40 µg/mL of the synthesized silver nanoparticles show an excellent zone of inhibition (Table **1**) against the four tested nosocomial pathogens. About 20 µg/mL of BSNPs revealed the best zone of inhibition against *E. faecalis* 14 mm ± 1.5 mm, *S. aureus* 11 mm ± 1.6 mm, and *P. aeuroginosa* 16 mm ± 1.3 mm, did not reveal the zone of inhibition for *K. Pneumoniae*. Similarly, 40 µg/mL of BSNPs had shown significant Zone of Inhibition for all three pathogens except *K. Pneumoniae* with lesser Zone of Inhibition. The researcher had also tested the efficacy of incorporated silver nanoparticles in the form of a spray, which suggests the possibility of reducing bacterial count to 50% and control of viral pathogens to 10%. This may be further validated by using infrared rays through an automated microbial load checker in the future, which will benefit the society as a whole.

## CONSENT FOR PUBLICATION

Not applicable.

## CONFLICT OF INTEREST

The authors confirm that this chapter contents have no conflict of interest.

## ACKNOWLEDGEMENTS

Declared none.

## REFERENCES

[1]     Anandalakshmi K, Venugobal J, Ramasamy V. Characterization of silver nanoparticles by green synthesis method using *pedalium murex* leaf extract and their antibacterial activity. Appl Nanosci 2016; 6(3): 399-408.
[http://dx.doi.org/10.1007/s13204-015-0449-z]

[2]     Jamil B, Hasan F, Hameed A, Ahmed S. Isolation of *bacillus subtilis* mk-4 from soil and its potential of polypeptide antibiotic production. Pak J Pharm Sci 2007; 20(1): 26-31.
[PMID: 17337424]

[3]     El Solh AA, Alhajhusain A. Update on the treatment of *Pseudomonas aeruginosa* pneumonia. J Antimicrob Chemother 2009; 64(2): 229-38.
[http://dx.doi.org/10.1093/jac/dkp201] [PMID: 19520717]

[4]     Haytham MM. Green synthesis and characterization of silver nanoparticles using banana peel extract and their antimicrobial activity against representative microorganisms. Journal of Radiation Research and Applied Sciences 2015; 111-9.

[5]     Kalishwaralal K, Deepak V, Ramkumarpandian S, Nellaiah H, Sangiliyandi G. Extracellular biosynthesis of silver nanoparticles by the culture supernatant of *bacillus licheniformis*. Mater Lett 2008; 62: 4411-3.
[http://dx.doi.org/10.1016/j.matlet.2008.06.051]

[6]     Lateef A, Ojo SA, Oladejo SM. Anti-candida, anti-coagulant and thrombolytic activities of biosynthesized silver nanoparticles using cell-free extract of Bacillus safensis LAU 13. Process Biochem 2016; 51(10): 1406-12.
[http://dx.doi.org/10.1016/j.procbio.2016.06.027]

[7]     Logan NA. Bacillus and relatives in foodborne illness. J Appl Microbiol 2012; 112(3): 417-29.
[http://dx.doi.org/10.1111/j.1365-2672.2011.05204.x] [PMID: 22121830]

[8]     Perrin H, *et al.* Paeni bacillus polymyxa produces fusaricidin-type antifungal antibiotics active against leptosphaeria maculans, the causative agent of blackleg disease of canola. Canada Journal of Microbial science 2002; 48: 159- 169.

[9]     Priyadarshini S, Gopinath V, Priyadharsshini NM, Ali DM, Velusamy P. Synthesis of anisotropic silver nanoparticles using novel strain, *Bacillus flexus* and its biomedical application. Colloids Surf B Biointerfaces 2013; 102: 232-7.

[http://dx.doi.org/10.1016/j.colsurfb.2012.08.018]

[10]    Rajesh S, Dharanishanthi V, Vinoth Kanna A. Antibacterial mechanism of biogenic silver nanoparticles of *lactobacillus acidophilus*. J Exp Nanosci 2014; 11: 124-9.

[11]    Sintubin L, De Gusseme B, Van der Meeren P, Pycke BF, Verstraete W, Boon N. The antibacterial activity of biogenic silver and its mode of action. Appl Microbiol Biotechnol 2011; 91(1): 153-62.
        [http://dx.doi.org/10.1007/s00253-011-3225-3] [PMID: 21468709]

[12]    Stein T. *Bacillus subtilis* antibiotics: structures, syntheses and specific functions. Mol Microbiol 2005; 56(4): 845-57.
        [http://dx.doi.org/10.1111/j.1365-2958.2005.04587.x] [PMID: 15853875]

[13]    Lledo W, Hernandez M, Lopez E, *et al.* Guidance for control of infections with carbapenem-resistant or carbapenemase-producing enterobacteriaceae in acute care facilities. MMWR 20 2009; 58(10): 256-60.

[14]    Bauer AW, Kirby WMM, Sherris JC, Turck M. Antibiotic susceptibility testing by a standardized single disk method. Am J Clin Pathol 1966; 45(4): 493-6.
        [http://dx.doi.org/10.1093/ajcp/45.4_ts.493] [PMID: 5325707]

# SUBJECT INDEX

power plant discharge 108
sewage 35
Whole genomic sequencing (WGS) 142
Wisconsin pollution discharge elimination
        System 108

## Z

Zeta analysis 180
Zooplankton 90

www.ingramcontent.com/pod-product-compliance
Lightning Source LLC
Chambersburg PA
CBHW050840220326
41598CB00006B/416